Springer Tracts in Advanced Robotics 85

For further volumes:
http://www.springer.com/series/5208

STAR (Springer Tracts in Advanced Robotics) has been promoted under the auspices of EURON (European Robotics Research Network)

Radu Bogdan Rusu

Semantic 3D Object Maps for Everyday Robot Manipulation

Springer

Author
Radu Bogdan Rusu
Open Perception, Inc.
San Francisco, CA
USA

ISSN 1610-7438 e-ISSN 1610-742X
ISBN 978-3-642-35478-6 e-ISBN 978-3-642-35479-3
DOI 10.1007/978-3-642-35479-3
Springer Heidelberg New York Dordrecht London

Library of Congress Control Number: 2012954153

Printed on acid-free paper

Springer is part of Springer Science+Business Media (www.springer.com)

Foreword

Robotics is undergoing a major transformation in scope and dimension. From a largely dominant industrial focus, robotics is rapidly expanding into human environments and vigorously engaged in its new challenges. Interacting with, assisting, serving, and exploring with humans, the emerging robots will increasingly touch people and their lives.

Beyond its impact on physical robots, the body of knowledge robotics has produced is revealing a much wider range of applications reaching across diverse research areas and scientific disciplines, such as: biomechanics, haptics, neurosciences, virtual simulation, animation, surgery, and sensor networks among others. In return, the challenges of the new emerging areas are proving an abundant source of stimulation and insights for the field of robotics. It is indeed at the intersection of disciplines that the most striking advances happen.

The *Springer Tracts in Advanced Robotics (STAR)* is devoted to bringing to the research community the latest advances in the robotics field on the basis of their significance and quality. Through a wide and timely dissemination of critical research developments in robotics, our objective with this series is to promote more exchanges and collaborations among the researchers in the community and contribute to further advancements in this rapidly growing field.

The monograph by Radu Bogdan Rusu is a contribution aimed at developing novel representations of robots' operating environments. The promising technology of point cloud-based perception is introduced to derive semantic 3D object models enabling complex mobile manipulation scenarios. Results are validated in a rich set of experiments on different robots performing a variety of tasks in dynamically changing indoor environments, revealing a promising outlook toward the goal of employing autonomous robots in everyday habitats.

Remarkably, the monograph is based on the author's doctoral thesis, which was a finalist for the Tenth Edition of the EURON Georges Giralt PhD Award devoted to the best PhD thesis in Robotics in Europe. A very fine addition to STAR!

Naples, Italy
October 2012

Bruno Siciliano
STAR Editor

Abstract

Environment models serve as important resources for an autonomous robot by providing it with the necessary task-relevant information about its habitat. Their use enables robots to perform their tasks more reliably, flexibly, and efficiently. As autonomous robotic platforms get more sophisticated manipulation capabilities, they also need more expressive and comprehensive environment models: for manipulation purposes their models have to include the objects present in the world, together with their position, form, and other aspects, as well as an interpretation of these objects with respect to the robot tasks.

This book proposes *Semantic 3D Object Models* as a novel representation of the robot's operating environment that satisfies these requirements and shows how these models can be automatically acquired from dense 3D range data. The book contributes in two important ways to the research area acquisition of environment models.

The first contribution is a novel *framework for Semantic 3D Object Model acquisition from Point Cloud Data*. The functionality of this framework includes robust alignment and integration mechanisms for partial data views, fast segmentation into regions based on local surface characteristics, and reliable object detection, categorization, and reconstruction. The computed models are *semantic* in that they infer structures in the data that are meaningful with respect to the robot task. Examples of such objects are doors and handles, supporting planes, cupboards, walls, or movable smaller objects.

The second key contribution is *point cloud representations based on 3D point feature histograms (3D-PFHs)*, which model the local surface geometry for each point. 3D-PFHs distinguish themselves from alternative 3D feature representations in that they are very fast to compute, robust against variations in pose and sampling density, and cope well with noisy sensor data. Their use substantially improves the quality of the *Semantic 3D Object Models* acquired, as well as the speed with which they are computed. 3D-PFHs come with specific software tools that allow for the learning of surface characteristics based on their underlying geometry, the assembly of most distinctive 3D points from a given cloud, as well as limited view-invariant correspondence search for 3D registration.

The contributions presented in this book have been fully implemented and empirically evaluated on different robots performing different tasks in different environments. The first demonstration relates to the problem of cleaning tables by disposing the objects on them into a garbage bin with a personal robotic assistant in the presence of humans in its working space. The framework for Semantic 3D Object Model acquisition is demonstrated and used to construct dynamic 3D collision maps, annotate the surrounding world with semantic labels, and extract object clusters supported by tables in real-time performance. The second demonstration presents an on-the-fly model acquisition system for door and handle identification from noisy 3D point cloud maps. Experimental results show good robustness in the presence of large variations in the data, without suffering from the classical under or over-fitting problems usually associated with similar initiatives based on machine learning classifiers. The third application example tackles the problem of real-time semantic mapping of indoor environments with different kinds of terrain classes, such as walkways and stairs, for the navigation of a six-legged robot with terrain-specific walking modes.

Contents

Part III Applications

List of Figures

List of Tables

List of Algorithms

List of Symbols and Notations

The list below contains the mathematical symbols and notations that are used most frequently throughout the book.

Symbol	Description
$\boldsymbol{p}_i$	a 3D point with $\{x_i, y_i, z_i\}$ geometric coordinates
$\langle \boldsymbol{p}_i, \boldsymbol{p}_j \rangle$	the dot product between $\boldsymbol{p}_i$ and $\boldsymbol{p}_j$
$\vec{\boldsymbol{n}}_i$	a surface normal estimate at a point $\boldsymbol{p}_i$ having a $\{n_{x_i}, n_{y_i}, n_{z_i}\}$ direction
$\mathcal{P}_n = \{\boldsymbol{p}_1, \boldsymbol{p}_2, \cdots\}$	a set of nD points (also represented by $\mathcal{P}$ for simplicity)
$\mathcal{P}^k$	the set of points $\boldsymbol{p}_j$, $(j \leq k)$, located in the k-neighborhood of a query point $\boldsymbol{p}_i$
r	the radius of a sphere for determining a point neighborhood $\mathcal{P}^k$
$\overline{\boldsymbol{p}}$	the mean (centroid) point of a set of points $\mathcal{P}$, $\overline{\boldsymbol{p}} = \frac{1}{k} \cdot \sum_{i=1}^{k} \boldsymbol{p}_i$
$\vec{z}$	the z-axis of a given coordinate system

Chapter 1
Introduction

"Our vision for robotics depends on perception..."

GARY BRADSKI

The population in Europe and many other countries is aging — probably causing the share of people aged 65 years or over to double within the next 40 years. As life expectancy increases, so does the chance of people becoming physically and cognitively limited or disabled, often leading to care-dependency. In the same way as the number of people requiring care is projected to increase is the number of people able to give care expected to decrease. This societal development forces us to develop new concepts, including technologies, for promoting independent living. Prolonging the independence of elderly people with minor disabilities and increasing their participation in daily life is expected to improve the well-being and the health-state of these people and thereby also mitigate the care-giving problem.

As a consequence, recent research programs as well as technology roadmaps propose personal assistive robots as a key technology for prolonging the independence of elderly people [Sch07]. Leading research centers in Japan [ItSMaUoT07], in the United States [Con09] and large cooperative research efforts in Europe [Pla09] are currently investigating technologies for the realization of mobile personal robots that can assist people in performing their daily activities. These envisioned personal assistive robots are to bring people their walking stick that they might have forgotten in another room the day before for example, and thereby increase the people safety. The robots might also help the people setting the table, loading the dishwasher, and cleaning up. Indeed a robot capable of performing pick-and-place tasks for the objects of daily use might already be of substantial use.

The realization of such personal assistive robots requires us to equip the robots with the necessary perceptual capabilities. The robots have to detect, recognize, localize, and geometrically reconstruct the objects in their environments in order to manipulate them competently. They have to interpret the sensor data they receive in the context of the actions and activities they perform. In order to get a cup out of the cupboard, the robot has to find the door handle to open the cupboard.

Fortunately, the robots do not have to solve these perceptual tasks everyday from anew. Because they are to perform many activities on a regular basis, because they are to manipulate the same objects over and over again, and because the environment

R.B. Rusu: *Semantic 3D Object Maps for Everyday Robot Manipulation*, STAR 85, pp. 1–11.
DOI: 10.1007/978-3-642-35479-3_1

is kept stable, for example by cleaning up, the robot can learn models of its environment and the objects it is to manipulate and use the learned models to simplify its tasks and in particular its perceptual tasks.

Environment models serve as important resources for an autonomous robot by providing it with the necessary task-relevant information about its habitat. Robots can make use of environment models such that they perform their tasks more reliably, flexibly, and efficiently. For example, a robot assistant that knows where cups are located, does not have to search for them. Similarly, a robot can only clean up a part of the world if it understands what are the objects that can be picked up but also the location where they normally belong. In general, robots must know information about the functions of objects such as the handles of doors, in order to manipulate them competently. To function efficiently, it is imperative that these environment models are acquired autonomously and therefore support the deployment of mobile robots in new environments, without requiring too many software updates or manual user input, because they are able to adapt themselves to the changes in the environment.

This dissertation book investigates the perceptual capabilities of personal assistive robots that can acquire models of their environments and the objects they are to manipulate.

The learning and use of environment models has a longstanding history in autonomous mobile robotics. Elfes [Elf89] pioneered the area of environment mapping and equipped robots with the means in order to localize themselves within the environment, or to determine what are their destinations and how to move from the current positions there efficiently. If these models would be inexistent, the robots would have to be programmed for each navigation task individually and they would need to perform exorbitant costly searches through the environment in order to find their respective destinations.

As autonomous robotic platforms get more sophisticated manipulation capabilities, they also need more expressive and comprehensive environment models: for manipulation purposes their models have to include the objects present in the world, information about how to grasp them, as well as an interpretation of these objects with respect to the robot tasks. In particular, in human living environments, the objects of interest are actually designed for being manipulated, for example: glasses are cylinders so that they can be grasped easily, mugs and pots have handles, while drawers, cupboards and kitchen appliances have fixtures for operating them.

More importantly, if a robot is asked to bring a glass it must not bring a dirty one or one that is intended for the use of somebody else. Generally put, perception for mobile manipulation must become a resource for the robot, which informs the robot with respect to what to do, to which object, and in which way. This represents one of the main issues of semantic perception for mobile manipulation.

1.1 Why "3D" Semantic Perception?

Humans perceive their environments in terms of images, and describe the world in terms of what they see. This erroneously suggests that we might be able to frame the perception problem solely in a 2D context. As we can see from Figures 1.1 and 1.2, framing the perception problem this way can lead to failures in capturing the true semantic meaning of the world.

The reasons are twofold. One one hand, monocular computer vision applications are flustered by both fundamental deficiencies in the current generation of camera devices and limitations in the datastream itself. The former will most likely be addressed in time, as technology progresses and better camera sensors are developed. An example of such a deficiency is shown in Figure 1.1, where due to the low dynamic range of the camera sensor, the right part of the image is completely underexposed. This makes it very hard for 2D image processing applications to recover the necessary information for recognizing objects in such scenes.

Fig. 1.1 Model matching failure in underexposed 2D images. None of the features extracted from the model (left) can be successfully matched onto the object of interest (right).

The second reason is linked directly to the fact that computer vision applications make use of 2D camera images mostly, which are inherently capturing only a projection of the 3D world. Figure 1.2 attempts to capture this problem by showing two examples of matching a certain object template in 2D camera images. While the system apparently matched the model template (a beer bottle in this case) successfully in the left part of the image, after we zoom out, we observe that the bottle of beer was in fact another picture in itself, stitched to a completely different 3D geometric surface (the body of a mug in this case). This is a clear example which shows that the semantics of a particular solution obtained only using 2D image processing can be lost if the geometry of the object is not incorporated in the reasoning process.

1.2 Computational Problems

The main computational problems of semantic perception that are investigated in this book are depicted in Figures 1.3 and 1.4. In both figures, the input data consists of range measurements coming from lasers scanners, thus the input is a set of points

Fig. 1.2 An example of a good model match using features extracted from 2D images (left), where a beer bottle template is successfully identified in an image. However, zooming out from the image, we observe that the bottle of beer is in fact another picture stitched to a completely different 3D object (in this case a mug). The semantics of the objects in this case are thus completely wrong.

with each point having a 3D coordinate in the world. Typically these points result from estimations of the reflectance of the laser beams from object surfaces. Though in most cases the points are close to the real scanned surfaces, they only represent rough approximations of them, and in some cases they are sampled behind or in front of the surface.

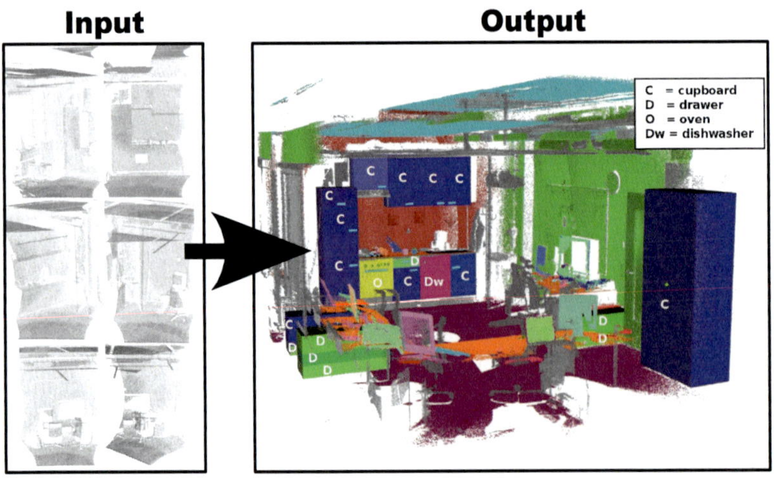

Fig. 1.3 Computational problem 1: create a Semantic 3D Object Map representation (right) for a kitchen environment from a set of input datasets (left)

The first computational problem that we want to investigate can be formulated as follows. Given a set of individual datasets as the ones presented in the left part of Figure 1.3, build a Semantic 3D Object Map such as the one presented in the right part of the figure, that contains information about the objects in the world such as: cupboards are cuboid containers with a frontal door and a handle, tables are supporting planes which can hold objects that are to be manipulated, and kitchen appliances contain knobs and handles that can be used to operate them.

The second computational problem is given in Figure 1.4. Given a dataset representing a supporting table plane with an object present on it (left), produce a representation such as the one shown on the right part, where the object model is decomposed into individual parts to help grasping applications and its surface is reconstructed using smooth triangular meshes.

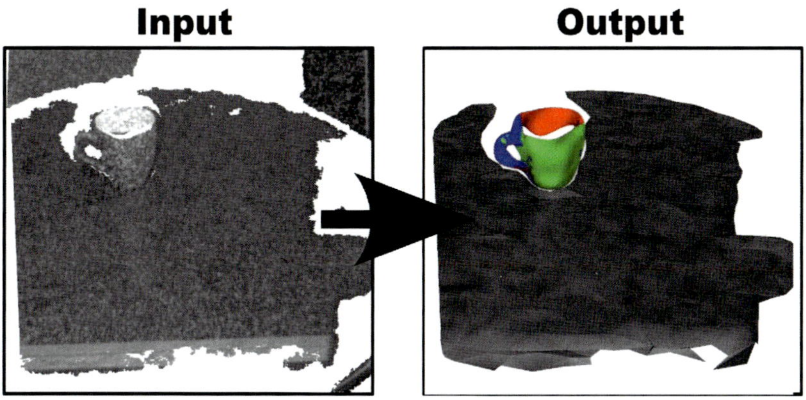

Fig. 1.4 Computational problem 2: decompose an object into individual parts and reconstruct its surface (right) from a raw dataset (left)

These two examples pose serious challenges for the current generation of 3D perception systems. Though both applications are formulated in the context of structured human living environments, the variation of object placements and shapes, as well as the differences from one room and apartment to another, makes the problem to be solved extremely difficult. In particular the acquired datasets contains exacerbated levels of noise, there are objects and surfaces which do not return measurements thus leaving holes in the data, the scenes are too cluttered and the objects to be searched for contain extremely large – if not infinite – variations of shapes and colors, just to mention a few.

1.3 Publications

The work presented in this book spawned a series of publications presented at major conferences and top journals in the field. Below is an excerpt from this list:

Journals

- Radu Bogdan Rusu, Jan Bandouch, Franziska Meier, Irfan Essa, Michael Beetz. *Human Action Recognition using Global Point Feature Histograms and Action Shapes.* Advanced Robotics journal, Robotics Society of Japan (RSJ), 2009.

- Radu Bogdan Rusu, Aravind Sundaresan, Benoit Morisset, Kris Hauser, Motilal Agrawal, Jean Claude Latombe, Michael Beetz. *Leaving Flatland: Efficient Real-Time 3D Navigation.* Journal of Field Robotics, special issue on 3D Mapping, 2009.
- Radu Bogdan Rusu, Zoltan Csaba Marton, Nico Blodow, Mihai Dolha, Michael Beetz. *Towards 3D Point Cloud Based Object Maps for Household Environments.* Robotics and Autonomous Systems Journal, special issue on Semantic Knowledge, 2008.
- Radu Bogdan Rusu, Brian Gerkey, Michael Beetz. *Robots in the kitchen: Exploiting ubiquitous sensing and actuation.* Robotics and Autonomous Systems Journal, special issue on Network Robot Systems, 2008.

Conferences

- Radu Bogdan Rusu, Andreas Holzbach, Rosen Diankov, Gary Bradski, Michael Beetz. *Perception for Mobile Manipulation and Grasping using Active Stereo.* Proceedings of the 9th IEEE-RAS International Conference on Humanoid Robots (Humanoids), Paris, France, December 7-10, 2009.
- Nico Blodow, Radu Bogdan Rusu, Zoltan Csaba Marton, Michael Beetz. *Partial View Modeling and Validation in 3D Laser Scans for Grasping.* Proceedings of the 9th IEEE-RAS International Conference on Humanoid Robots (Humanoids), Paris, France, December 7-10, 2009.
- Ulrich Klank, Dejan Pangercic, Radu Bogdan Rusu, Michael Beetz. *Real-time CAD Model Matching for Mobile Manipulation and Grasping.* Proceedings of the 9th IEEE-RAS International Conference on Humanoid Robots (Humanoids), Paris, France, December 7-10, 2009.
- Zoltan Csaba Marton, Lucian Goron, Radu Bogdan Rusu, Michael Beetz. *Reconstruction and Verification of 3D Object Models for Grasping.* Proceedings of the 14th International Symposium on Robotics Research (ISRR), Lucernce, Switzerland, August 31 - September 3 2009.
- Radu Bogdan Rusu, Wim Meeussen, Sachin Chitta, M. Beetz. *Laser-based Perception for Door and Handle Identification.* Proceedings of International Conference on Advanced Robotics (ICAR), Munich, Germany, June 22-26 2009. **Best paper award**.
- Radu Bogdan Rusu, Ioan Alexandru Sucan, Brian Gerkey, Sachin Chitta, Michael Beetz, Lydia E. Kavraki. *Real-time Perception-Guided Motion Planning for a Personal Robot.* Proceedings of IEEE/RSJ International Conference on Intelligent Robots and Systems (IROS), St. Louis, MO, USA, October 11-15 2009.
- Zoltan Csaba Marton, Radu Bogdan Rusu, Dominik Jain, Uli Klank, Michael Beetz. *Probabilistic Categorization of Kitchen Objects in Table Settings with a Composite Sensor.* Proceedings of IEEE/RSJ International Conference on Intelligent Robots and Systems (IROS), St. Louis, MO, USA, October 11-15 2009.

- Radu Bogdan Rusu, Zoltan Csaba Marton, Nico Blodow, Andreas Holzbach, Michael Beetz. *Model-based and Learned Semantic Object Labeling in 3D Point Cloud Maps of Kitchen Environments.* Proceedings of IEEE/RSJ International Conference on Intelligent Robots and Systems (IROS), St. Louis, MO, USA, October 11-15 2009.
- Radu Bogdan Rusu, Nico Blodow, Zoltan Csaba Marton, Michael Beetz. *Close-range Scene Segmentation and Reconstruction of 3D Point Cloud Maps for Mobile Manipulation in Human Environments.* Proceedings of IEEE/RSJ International Conference on Intelligent Robots and Systems (IROS), St. Louis, MO, USA, October 11-15 2009.
- Radu Bogdan Rusu, Andreas Holzbach, Nico Blodow, Michael Beetz. *Fast Geometric Point Labeling using Conditional Random Fields.* Proceedings of IEEE/RSJ International Conference on Intelligent Robots and Systems (IROS), St. Louis, MO, USA, October 11-15 2009.
- Radu Bogdan Rusu, Nico Blodow, Michael Beetz. *Fast Point Feature Histograms (FPFH) for 3D Registration.* Proceedings of IEEE International Conference on Robotics and Automation (ICRA), Kobe, Japan, May 12-17 2009.
- Zoltan Csaba Marton, Radu Bogdan Rusu, Michael Beetz. *On Fast Surface Reconstruction Methods for Large and Noisy Datasets.* Proceedings of IEEE International Conference on Robotics and Automation (ICRA), Kobe, Japan, May 12-17 2009.
- Benoit Morisset, Radu Bogdan Rusu, Aravind Sundaresan, Kris Hauser, Motilal Agrawal, Jean Claude Latombe, Michael Beetz. *Leaving Flatland: Toward Real-Time 3D Navigation.* Proceedings of IEEE International Conference on Robotics and Automation (ICRA), Kobe, Japan, May 12-17 2009.
- Michael Beetz, Freek Stulp, Bernd Radig, Jan Bandouch, Nico Blodow, Mihai E. Dolha, Andreas Fedrizzi, Dominik Jain, Uli Klank, Ingo Kresse, Alexis Maldonado, Zoltan Csaba Marton, Lorenz Moesenlechner, Federico Ruiz, Radu Bogdan Rusu, Moritz Tenorth. *Cognition, Control and Learning for Everyday Manipulation Tasks in Human Environments.* Proceedings of IEEE 17th International Symposium on Robot and Human Interactive Communication (RO-MAN), Munich, Germany, August 1-3 2008. **Invited paper**.
- Radu Bogdan Rusu, Aravind Sundaresan, Benoit Morisset, Motilal Agrawal, Michael Beetz. *Leaving Flatland: Realtime 3D Stereo Semantic Reconstruction.* Proceedings of International Conference on Intelligent Robotics and Applications (ICIRA), Wuhan, China, October 15-17 2008.
- Radu Bogdan Rusu, Zoltan Csaba Marton, Nico Blodow, Michael Beetz. *Learning Informative Point Classes for the Acquisition of Object Model Maps.* Proceedings of 10th International Conference on Control, Automation, Robotics and Vision (ICARCV), Hanoi, Vietnam, December 17-20 2008.
- Radu Bogdan Rusu, Zoltan Csaba Marton, Nico Blodow, Mihai Dolha, Michael Beetz. *Functional Object Mapping of Kitchen Environments.* Proceedings of 21st IEEE/RSJ International Conference on Intelligent Robots and Systems (IROS), Nice, France, September 22-26 2008.

- Radu Bogdan Rusu, Nico Blodow, Zoltan Csaba Marton, Michael Beetz. *Aligning Point Cloud Views using Persistent Feature Histograms.* Proceedings of 21st IEEE/RSJ International Conference on Intelligent Robots and Systems (IROS), Nice, France, September 22-26 2008.
- Radu Bogdan Rusu, Zoltan Csaba Marton, Nico Blodow, Michael Beetz. *Persistent Point Feature Histograms for 3D Point Clouds.* Proceedings of 10th International Conference on Intelligent Autonomous Systems (IAS-10), Baden-Baden, Germany, July 23-25 2008.
- Radu Bogdan Rusu, Jan Bandouch, Zoltan Csaba Marton, Nico Blodow, Michael Beetz. *Action Recognition in Intelligent Environments using Point Cloud Features Extracted from Silhouette Sequences.* Proceedings of 17th IEEE International Symposium on Robot and Human Interactive Communication (RO-MAN), Munich, Germany, August 1-3 2008.
- Radu Bogdan Rusu, Nico Blodow, Zoltan Marton, Alina Soos, Michael Beetz. *Towards 3D Object Maps for Autonomous Household Robots.* Proceedings of the 20th IEEE International Conference on Intelligent Robots and Systems (IROS), San Diego, CA, USA, Oct 29 - 2 Nov 2007.
- Matthias Kranz, Alexis Maldonado, Radu Bogdan Rusu, Bernhard Hoernler, Gerhard Rigoll, Michael Beetz, Albrecht Schmidt. *Sensing Technologies and the Player Middleware for Context-Awareness in Kitchen Environments.* Proceedings of 4th International Conference on Networked Sensing Systems, Braunschweig, Germany, June 6-8 2007.

Workshops

- Michael Beetz, Nico Blodow, Ulrich Klank, Zoltan Csaba Marton, Dejan Pangercic, Radu Bogdan Rusu. *CoP-Man – Perception for Mobile Pick-and-Place in Human Living Environments.* Proceedings of the 22nd IEEE/RSJ International Conference on Intelligent Robots and Systems (IROS), workshop on Semantic Perception for Mobile Manipulation, St. Louis, USA, October 15, 2009. **Invited paper**.
- Radu Bogdan Rusu, Andreas Holzbach, Gary Bradski, Michael Beetz. *Detecting and Segmenting Objects for Mobile Manipulation.* Proceedings of IEEE Workshop on Search in 3D and Video (S3DV), held in conjunction with the 12th IEEE International Conference on Computer Vision (ICCV), Kyoto, Japan, 27 September, 2009.
- Zoltan Csaba Marton, Nico Blodow, Mihai Dolha, Moritz Tenorth, Radu Bogdan Rusu, Michael Beetz. *Autonomous Mapping of Kitchen Environments and Applications.* Proceedings of 1st International Workshop on Cognition for Technical Systems, Munich, Germany, October 6-8 2008.
- Radu Bogdan Rusu, Zoltan Csaba Marton, Nico Blodow, Michael Beetz. *Interpretation of Urban Scenes based on Geometric Features.* Proceedings of 21st IEEE/RSJ International Conference on Intelligent Robots and Systems (IROS) Workshop on 3D Mapping, Nice, France, September 26 2008. **Invited paper**.

- Radu Bogdan Rusu, Aravind Sundaresan, Benoit Morisset, Motilal Agrawal, Michael Beetz, Kurt Konolige. *Realtime Extended 3D Reconstruction from Stereo for Navigation.* Proceedings of 21st IEEE/RSJ International Conference on Intelligent Robots and Systems (IROS) Workshop on 3D Mapping, Nice, France, September 26 2008. **Invited paper.**
- Radu Bogdan Rusu, Alexis Maldonado, Michael Beetz, Brian Gerkey. *Extending Player / Stage / Gazebo towards Cognitive Robots Acting in Ubiquitous Sensor-equipped Environments.* Proceedings of IEEE International Conference on Robotics and Automation (ICRA) Workshop for Network Robot Systems, Rome, Italy, April 14 2007.
- Michael Beetz, Jan Bandouch, Alexandra Kirsch, Alexis Maldonado, Armin Müller, Radu Bogdan Rusu. *The Assistive Kitchen - A Demonstration Scenario for Cognitive Technical Systems.* Proceedings of 4th COE Workshop on Human Adaptive Mechatronics (HAM), 2007.

1.4 Book Outline and Contributions

Figure 1.5 presents a graphical roadmap depicting the organization of the book. The contributions of this book consist in numerous theoretical formulations and practical solutions to the problem of 3D semantic mapping with mobile robots. Divided into three distinct parts, the book starts by presenting important contributions to a set of algorithmic steps comprising the kernel of a Semantic 3D Object Mapping system. This includes dedicated conceptual architectures and their software implementations for 3D mapping systems, useful in solving complex perception problems. Starting from point-based representations of the world as preliminary data sources, an innovative solution for the characterization of the surface geometry around a point is given through the formulation of a Point Feature Histogram (PFH) 3D feature. The proposed feature space is extremely useful for the problem of finding point correspondences in multiple overlapping datasets, as well as for learning classes of geometric primitives that label point sampled surfaces. Other important contributions are presented in the subsequent chapters, by addressing the problems of partial view point cloud registration under geometric constraints, and surface segmentation and reconstruction from point cloud data for static environments.

The second part of the book tackles the semantic scene interpretation of indoor environments, including both methods based on machine learning classifiers and parametric shape model fitting, to decompose the environment into meaningful semantic objects useful for mobile manipulation scenarios. The former is formulated in the context of supervised learning, where multiple training examples are used to estimate the required optimal parameters for the classification of 3D points based on the underlying surface that they represent. The latter includes hierarchical non-linear geometric shape fitting using robust estimators and spatial decomposition techniques, capable of decomposing objects into primitive 3D geometries and creating hybrid shape-surface object models. To obtain informative classes of complete objects, a novel global feature estimator which captures the geometry of the

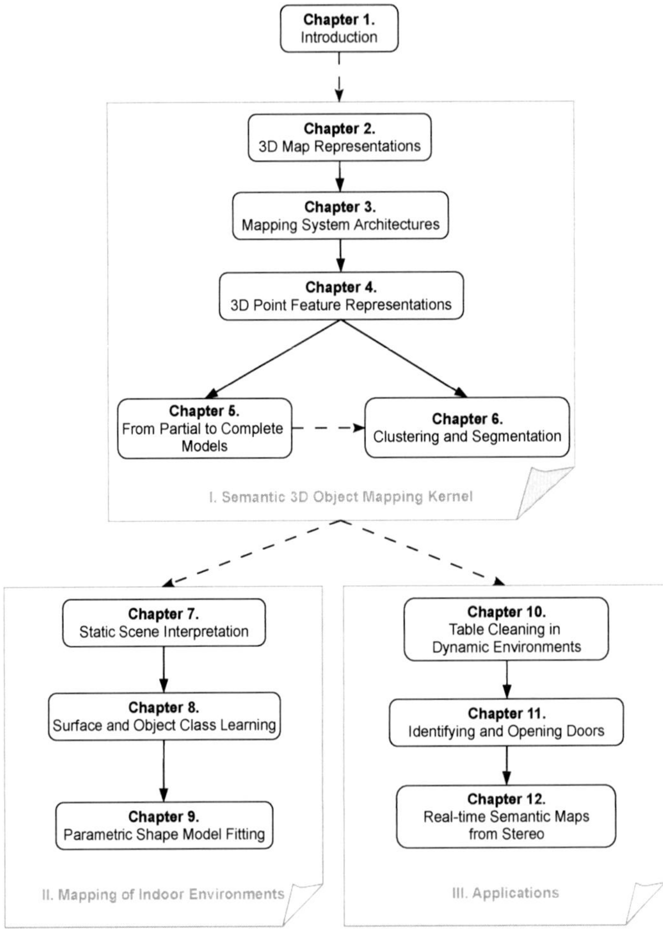

Fig. 1.5 A roadmap of the book. The source chapter of an arrow should be read before the destination chapter. Dashed arrows indicate optional ordering.

object using pairwise relationships between its surface components is proposed and validated on datasets acquired using active stereo cameras.

Finally, the third part of the book presents applications of comprehensive 3D perception systems on complete robotic platforms. The first application relates to the problem of cleaning tables by moving the objects present on them in the context of dynamic environments with personal robotic assistants. The proposed architecture constructs 3D dynamic collision maps, annotates the surrounding world with semantic labels, and extracts object clusters supported by tables with real-time performance. The experiments are validated on the PR2 (Personal Robotics 2) platform from Willow Garage, using the ROS (Robot Operating System) paradigms.

The second application uses the same mobile manipulation platform from Willow Garage, and presents an architecture for door and handle identification from noisy 3D point cloud maps. Experimental results show good robustness in the presence of large variations in the data, without suffering from the classical under or over-fitting problems usually associated with similar initiatives based on machine learning classifiers. The third and final complete application example tackles the problem of real-time semantic mapping from stereo data for navigation using the RHex (Robot Hexapod) mobile robot from Boston Dynamics.

Part I

Semantic 3D Object Mapping Kernel

Chapter 2
3D Map Representations

"PICTURE, n. A representation in two dimensions of something wearisome in three."

AMBROSE BIERCE (1842 - 1914?)

The most primitive data representation unit in a three-dimensional Euclidean space is the 3D point itself, $\boldsymbol{p}_i$. This chapter gives a basic overview on acquisition techniques for 3D points, as well as on representation formats for collections of points in the context of mapping with mobile robots. Figure 2.1 presents examples of very accurate collections of 3D points representing objects found in indoor environments.

2.1 Data Acquisition

By definition, we will refer to a collection of 3D points as a *point cloud* structure $\mathcal{P}$. Point clouds represent the basic input data format for 3D perception systems, and provide discrete, but meaningful representations of the surrounding world. Without any loss of generality, the $\{x_i, y_i, z_i\}$ coordinates of any point $\boldsymbol{p}_i \in \mathcal{P}$ are given with respect to a fixed coordinate system, usually having its origin at the sensing device used to acquire the data. This means that each point $\boldsymbol{p}_i$ represents the distance on the three defined coordinate axes from the acquisition viewpoint to the surface that the point has been sampled on. Though there are many ways of measuring distances and converting them to 3D points, in the context of mobile robotic applications [Nue09], the two most used approaches are:

- Time-of-Flight (TOF) systems, which measure the delay until an emitted signal hits a surface and returns to the receiver, thus estimating the true distance from the sensor to the surface;
- Triangulation techniques, which estimate distances by means of connecting correspondences seen by two different sensors at the same time. To compute the distance to a surface, the two sensors need to be *calibrated* with respect to each other, that is, their intrinsic and extrinsic properties must be known.

R.B. Rusu: *Semantic 3D Object Maps for Everyday Robot Manipulation*, STAR 85, pp. 15–25.
DOI: 10.1007/978-3-642-35479-3_2

Fig. 2.1 Example of very accurate point cloud datasets representing objects found in kitchen environments, acquired using structured light sensing devices

In addition, a special category of sensors use structured light to obtain high precision point cloud datasets such as the ones presented in Figure 2.1. However, with few exceptions, these sensing devices are not used in robotic applications in indoor environments, and their use is mostly limited to 3D modeling applications in special constrained environments.

The first category of systems presented above involves sensing devices such as a Laser Measurement System (LMS) or LIDAR, radars, Time-of-Flight (TOF) cameras, or sonar sensors, which send "rays" of light (*e.g.* laser) or sound (*e.g.* sonar) in the world, which will reflect and return to the sensor. Knowing the speed with which a ray propagates, and using precise circuitry to measure the exact time when the ray was emitted and when the signal returned, the distance d can be estimated as (simplified):

$$d = \frac{ct}{2} \tag{2.1}$$

where c represents the speed of the ray (*e.g.* speed of light for laser sensors), and t is the amount of time spent from when the ray was emitted until it was received back. In contrast, triangulation techniques usually estimate distances using the following equation (simplified):

$$d = \frac{fT}{\|x_1 - x_2\|} \tag{2.2}$$

where f represents the focal distance of both sensors, T the distance between the sensors, and x_1 and x_2 are the corresponding points in the two sensors. Though many different triangulation systems exist, the most popular system used in mobile robotics applications is the stereo camera. Figure 2.2 presents some of the sensing devices used for the acquisition of point cloud data in this book.

The laser measurement systems presented herein are inherently 2D, in the sense that they combine a laser range finder with a rotating mirror to measure distances on a plane. To obtain a 3D point cloud, they are placed on rotating units such as pan/tilt

Fig. 2.2 Example of perception sources used to generate 3D point cloud datasets

or robot arms as seen in the left part of Figure 2.2. Using the kinematics of these units, we can obtain a multitude of such 2D measurement planes, which can then be converted into a consistent 3D representation. One example of a point cloud $\mathcal{P}$ obtained using this principle is shown in Figure 2.3.

Fig. 2.3 An example of a point cloud dataset acquired from a 2D laser moved with an actuated tilting unit

While the resultant point cloud is accurate enough to be used for modeling applications, systems employing a rotating 2D laser are suffering from a few disadvantages, a main one being the data acquisition speed. Though the work in this book makes use of one of the fastest 2D laser systems to date, the SICK LMS 400, with acquisition frequencies of up to 500Hz, building a rotating unit that would move the entire laser unit with similar speeds in order to obtain 3D point clouds is extremely hard and could pose real dangers in the context of robotic systems working in indoor environments where people are also present.

A solution that can be employed to obtain dense 3D data faster, is to use Time-of-Flight camera systems, such as the SwissRanger 4000 (see Figure 2.2 right). These systems can provide 3D data representing a certain part of the world at frequencies up to 30Hz, thus enabling their usage in applications with fast reactive constraints. The resultant data is however noisier than the one acquired using laser sensors, not as dense (as the resolution of most TOF cameras is very low), and can suffer from big veiling effects [MDH$^+$08]. An example of such point cloud data, acquired using a SR 4000 TOF camera is shown in Figure 2.4.

Fig. 2.4 An example of a point cloud dataset acquired using a Time-of-Flight camera

The second category of sensors used in this book comprises stereo cameras. Besides acquisition speed, the major advantage of stereo cameras over systems working on a Time-of-Flight principle, is that they are *passive*, in the sense that they do not need to project or send any light or sound sources into the world in order to estimate distances. Instead, they just need to find point correspondences in both cameras that match a certain criterion. Without going into their working principle [HZ04] or implementation details [Kon97], it suffices to say that due to the correspondence problem, it is impossible to estimate good depth measurements for all data points (*e.g.* pixels) in the camera, especially in the absence of texture. Though in all fairness, better stereo matching techniques exist (see [NSG08] for a recent review), they are not yet suitable for robotic applications demanding fast update rates. This means that the resultant point cloud datasets acquired using passive stereo cameras are not always very dense, or *complete*, and might contain holes for portions of the scene where texture is absent or point correspondences are hard to estimate.

Figure 2.5 shows two different datasets acquired using a stereo system: a) in an environment where almost all surfaces are highly textured (left) and b) in an environment where certain surfaces have texture-less regions.

Fig. 2.5 Examples of point cloud datasets acquired using a stereo camera: in a textured environment (left); in an environment that lacks texture information in certain regions (right)

A solution in this sense is to make sure that the world contains only objects which exhibit good texture characteristics, or alternatively project random texture patterns onto the scene in order to improve the correspondence problem. The latter falls into the category of transforming the passive stereo approach into an active one, where intermittent flashes from texture generators could be projected on objects using a structured light projector. An example of the differences in the resultant depth images with and without such a projective texture approach are presented in Figure 2.6.

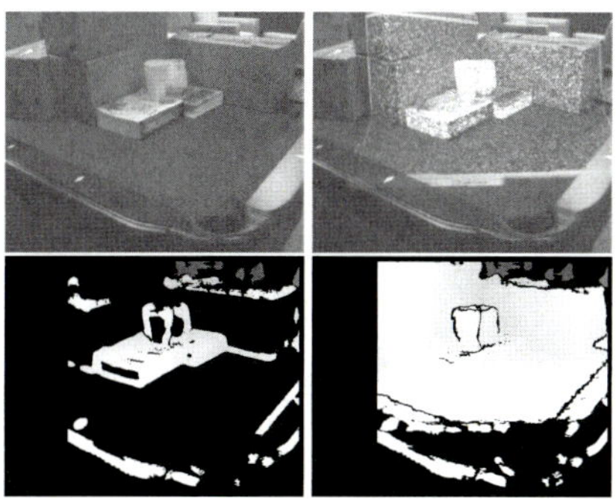

Fig. 2.6 An example of active stereo point cloud acquisition. Textured light in the right column yields a denser disparity map for stereo.

The sensing devices presented in the right part of Figure 2.2 present an example of a composite device that can potentially combine different data sources and fuse them together for the purpose of using the strengths of each device [MRJ+09]. Figure 2.7 presents a few example images acquired using the individual sensing devices which are part of the composite sensor system.

Fig. 2.7 Different images taken using the composite sensor device from Figure 2.2. From left to right: stereo camera left and right images, TOF camera depth image, and finally a thermal camera image.

2.2 Data Representation

Once a 3D point cloud dataset has been acquired using one of the methods presented in the previous section, it needs to go through a series of geometric processing steps in order to extract meaningful information that can help a robot in performing its tasks. It is the role of a mapping system therefore to process and convert the raw input point cloud data into different representations and formats based on the requirements imposed by each individual processing step.

An important aspect when dealing with point cloud representations, is that they are able to store much more information than just the 3D positions of points as acquired from the input sensing device. For example, if the data has been acquired using a 2D laser sensor, each measurement is usually accompanied by a surface remission value, which quantifies the intensity of the reflected laser beam. Figure 2.8 presents an example of a point cloud dataset acquired using a tilting 2D laser sensor, shown in remission scale using monochrome colors. The artifacts in the figure can be ignored, as all data is unprocessed (raw) and presented as acquired from the sensing device.

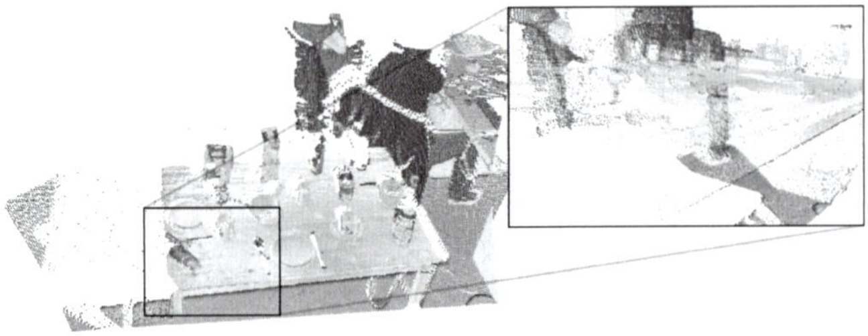

Fig. 2.8 Point Cloud Dataset showed in surface remission (grayscale) spectrum

Alternatively, during the acquisition process of a point cloud $\mathcal{P}$, the distances from the sensor to the surfaces in the world can be saved as properties for each resultant 3D point $\boldsymbol{p}_i \in \mathcal{P}$. Figure 2.9 represents the same point cloud dataset but this time colored in terms of the distance of the objects with respect to the sensor acquisition origin.

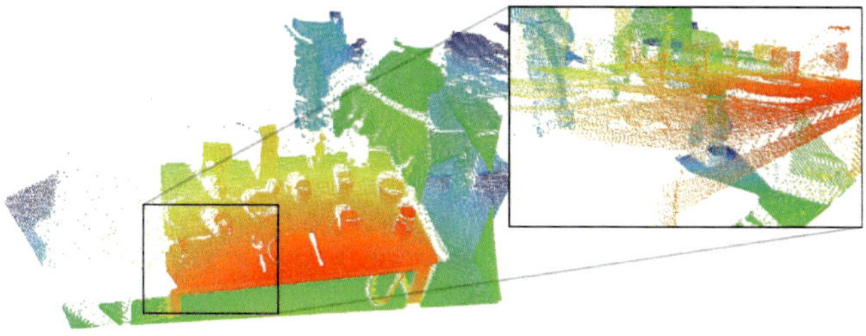

Fig. 2.9 Point Cloud Dataset showed in distance (red is close, blue is far away) spectrum

Processing steps such as point cloud segmentation and clustering can work directly on the raw point cloud data, and provide additional properties as outputs for each point. For example, in the context of an application that requires that supporting table planes are segmented and marked in the point cloud, and object clusters lying on these planes are clustered in an Euclidean sense (see Chapter 6), each point $\boldsymbol{p}_i \in \mathcal{P}$ will receive an additional property that takes the values of the new assigned class. Figure 2.10 presents an example of point annotations and colors each point based on the class label received, such as tables with light green, object clusters supported by them with a random color, and all remaining points are marked with black.

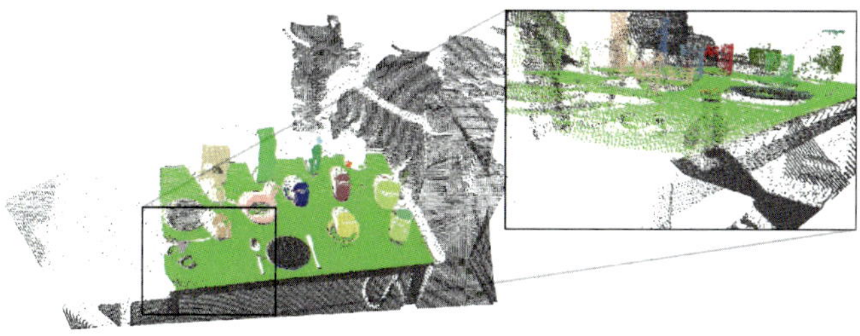

Fig. 2.10 Regions of interest in a point cloud dataset, represented using different class labels and colors. The colors are chosen randomly in the two separate images for the purpose of depicting the different object clusters and the table plane.

Giving the fact that all the above examples need point cloud representations which hold multiple properties per point, the definition of a point $\boldsymbol{p}_i = \{x_i, y_i, z_i\}$ changes to that of $\boldsymbol{p}_i = \{f_1, f_2, f_3 \cdots f_n\}$, where f_i defines a feature value in a given space (color, class label, geometry, etc), thus changing the concept of a 3D point to a nD one. From these requirements, we can deduce that an appropriate I/O data storage format for a point cloud $\mathcal{P}$, would be to save each point with all its attribute values on a new line in a file, and thus have a file with n lines for the n total number of points in $\mathcal{P}$. A fictitious example is shown below:

$$\begin{bmatrix} x_1 & y_1 & z_1 & r_1 & g_1 & b_1 & l_1 & d_1 \cdots \\ x_2 & y_2 & z_2 & r_2 & g_2 & b_2 & l_2 & d_2 \cdots \\ & & & & & & & \cdots \\ x_n & y_n & z_n & r_n & g_n & b_n & l_n & d_n \cdots \end{bmatrix}$$

where x_i, y_i, z_i represent the 3D point coordinates, r_i, g_i, b_i a color associated with each point, l_i a class label, and d_i the distance from the sensor to the surface. From an implementation point of view, the data could be stored as ASCII files, or it could alternatively be transformed into a binary representation, where each feature value has a fixed size (*e.g.* 32-bits). The latter could provide significant advantages in terms of loading and saving large point cloud datasets to disk, by making use of memory *mapping* mechanisms.

To understand the geometry around a query point $\boldsymbol{p}_q$, most geometric processing steps need to discover a collection of neighboring points $\mathcal{P}^k$ that represent the underlying scanned surface through sampling approximations. The mapping system therefore needs to employ mechanisms for enabling the search of point neighbors in fast ways, without re-computing distances between each other every time. A solution is to use spatial decomposition techniques such as kd-trees or octrees, and partition the point cloud data $\mathcal{P}$ into chunks, such that queries with respect to the location of the points in $\mathcal{P}$ can be answered fast. Though different from an implementation point of view, most spatial decomposition techniques can construct and give hints of a *volumetric* representation for a cloud $\mathcal{P}$, by enclosing all its points in boxes (also called "voxels") with different widths. An example of such representations is given in Figures 2.11 and 2.12 for octree data structures and box-decomposition (BD) trees respectively.

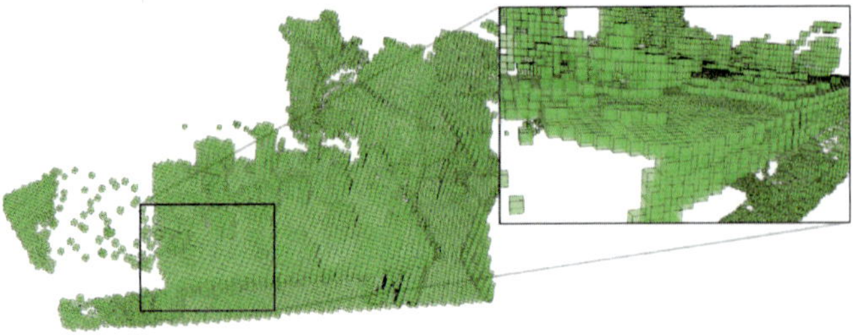

Fig. 2.11 Volumetric representation using an octree structure, with a leaf size of 1.5cm

Besides providing fast access to point locations and their corresponding neighbors, octree representations are also popular in the context of collision detection applications, where instead of estimating distances to each point, we can perform raycasting to the voxels encompassing $\mathcal{P}$ to discover the portions of space which are free or occluded. A big advantage of octree data structures is that they are easy to update and support point insertion and deletion almost natively.

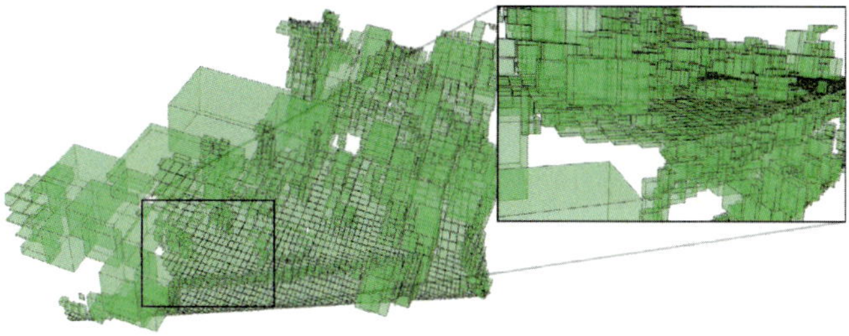

Fig. 2.12 Volumetric representation using a box decomposition tree (bd-tree) structure, with a bucket size of 30 points

Chapter 12 describes the implementation of a system that uses dynamic octree structures for the purpose of real-time mapping with a mobile robot.

The other spatial decomposition technique mentioned above, the bd-tree, represents one variant of a kd-tree structure optimized to provide a greater robustness for highly cluttered point cloud datasets [AMN$^+$98]. Its usage concerns mainly the fast search of approximate nearest neighbors solutions, as shown in Chapter 4. However, in contrast to octree structures, bd-trees or kd-trees are more difficult to update, and thus their usage is mostly limited to static scenes for applications working with individual point cloud datasets.

Mobile manipulation and grasping applications also require data representations that can approximate the surface better (*i.e.* in more detail) than the spatial decomposition representations shown above. When estimating grasping strategies or grasping points for a certain object, the contact forces between the end effector (*e.g.* gripper, hand) of the mobile robot and the object to be grasped have to be estimated.

In the context of this book we propose a series of surface reconstruction techniques, each well suited for particular applications. Figure 2.13 presents the classical triangular surface reconstruction approach for the same point cloud dataset presented above, using techniques from our previous work [MRB09]. The triangle vertices are represented by the points $\boldsymbol{p}_i \in \mathcal{P}$, while the edges can be stored in the following form:

$$\begin{bmatrix} v_{i_1} & v_{j_1} & v_{k_1} \\ v_{i_2} & v_{j_2} & v_{k_2} \\ & & \cdots \\ v_{i_m} & v_{j_m} & v_{k_m} \end{bmatrix}$$

where v_i, v_j, v_k represent point indices for $\boldsymbol{p}_i \in \mathcal{P}$, and m represents the total number of triangles created. This representation allows the combination of triangle meshes with other point cloud data attributes stored per point as mentioned above.

One disadvantage of triangular mesh representations is that they are costly to update if the scene changes. Taking for example the dataset in Figure 2.13, all objects located on the table are connected to it to form a compact and uniform surface representation. If one of the objects is removed from the table in a subsequent step however, the entire surface of the table might need to be regenerated, depending

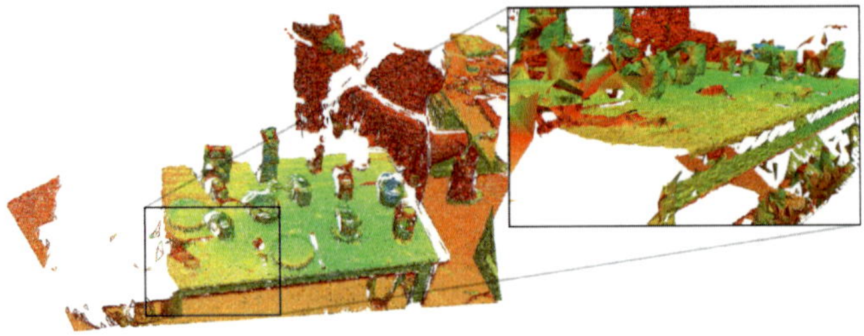

Fig. 2.13 Triangle meshes approximating the underlying sampled surface, showed in intensity (red shades) spectrum

on how many of the surface components are actually connected with each other. Though clever implementations can be written to account for this problem, a solution which gives good results in most cases is to model the surface using convex planar polygons, such as the ones presented in Figure 2.14. Here, the system models the point cloud dataset $\mathcal{P}$ first using an octree implementation with a small leaf width (see Chapter 12 for a comprehensive example), and then proceeds at estimating sets of planar patches that best approximate the underlying surface model. The resultant models can be more inaccurate than triangular surface meshes, but they provide important advantages in terms of the update speed for applications with tight computational constraints.

Fig. 2.14 Convex planar patches approximating the underlying sampled surface

Another interesting solution is to segment primitive geometric surfaces in $\mathcal{P}$ for the purpose of providing smoother data approximations, using 3D shapes such as cylinders, planes, spheres, cones, etc. For each object cluster such as the ones presented in Figure 2.10, the system can first attempt to fit a primitive geometric shape, then use triangular meshes to model the remaining points. The rest of the data in $\mathcal{P}$ can be left as points and labeled as such, thus providing a true *hybrid* representation of the scene, using a multitude of formats presented so far: points, geometric models, and triangular meshes.

Fig. 2.15 Hybrid Point-Surface-Models representations, created by fitting 3D geometric primitives to the point cloud dataset, and triangulating residual cluster points. The colors are chosen randomly in the two separate images for the purpose of depicting the different object clusters and the table plane.

2.3 Summary

This chapter presented basic acquisition schemes for 3D point cloud datasets in the context of applications with mobile robots in indoor environments. Once an input point cloud dataset is acquired, it needs to go through a series of processing steps, which require or store the output in various different data representation formats. In general, point clouds representations are easy to work with, as they represent the raw scanned data. However, they require large quantities of memory and their large size makes certain geometric processing operations very difficult. Spatial decomposition structures alleviate both constraints, and provide simple and fast access to the underlying points by the use of efficient tree structures.

Representations that approximate the underlying scanned surface with convex polygonal patches, or triangular meshes, are useful for collision detection applications and in general mobile grasping where the contact forces between the end effector (*e.g.* gripper, hand) of the mobile robot and the object to be grasped have to be estimated.

Chapter 3
Mapping System Architectures

"Each new situation requires a new architecture."

JEAN NOUVEL

To use a well known metaphor, designing the architecture of a 3D perception system can be anything from a walk in the park to a day of fishing. This chapter takes on the challenge of proposing a comprehensive system architecture for the Semantic 3D Object Mapping kernel used in this book. Given the scenario laid out in the introduction, we are investigating the following computational problem: from a set of point cloud representations, each resulting from 3D laser scans acquired in an indoor environment, say a kitchen, automatically compute an integrated semantic representation of the data, with cupboards and drawers represented as cuboid containers with front doors and handles, tables and shelves as horizontal planes, and objects of interest classified and categorized based on their properties and the underlying geometric surface that they represent. Since this represents nothing else but the "holy grail" of 3D mapping systems from a data interpretation point of view (in the context of our application scenario), the resultant system architecture would need to comprise a big list of geometric and learning steps, where the output of a step requires the input of another, and so on.

Though designing a complex system architecture on paper is easy, implementing it on a mobile manipulation platform with constrained resources is a different story. Thus before adventuring at proposing any particular architecture at all, we need to first build a list of requirements that such a complex system needs to fulfill in order to function efficiently, robustly, and for long periods of time without interruptions:

- *Modularity*. Though the type of algorithmic steps could be set beforehand for general purpose tasks such as: registration - align partial data together in the same frame, surface reconstruction - create a surface representation from sampled point data, or region segmentation - determine regions of interest in the data and mark them appropriately, the architecture should allow the replacement of algorithms in one class of problems with different or more efficient implementations when they become available.
- *Robustness*. Assumptions referring to the results obtained from a particular algorithm should not be absolute, in the sense that errors or mistakes should be

R.B. Rusu: *Semantic 3D Object Maps for Everyday Robot Manipulation*, STAR 85, pp. 27–31.
DOI: 10.1007/978-3-642-35479-3_3

tolerated and dealt with accordingly. Though it becomes difficult to formulate this in a general sense, one particular example is represented by segmentation results obtained from a planar decomposition step for a dataset in a region growing framework (see Chapter 7). Though in a first step not all regions are detected properly, the system falls back to a secondary re-segmentation of the areas using the information acquired so far and a set of fixtures (handles and knobs) detected in the vicinity of these planar areas.

- *Efficiency*. All methods should be designed from the beginning to run on systems with tight computational constraints (*i.e.* mobile manipulation platforms). While measuring the efficiency of a particular algorithm, it must be noted that under realistic situations, some of the computational resources usually available could be used by other subsystems (*e.g.* motion planning, grasping, navigation).
- *Different levels of detail*. Related to the previously mentioned step, in the context of the availability of computational resources, the mapping architecture must be able to provide intermediate but useful results to its consumers. For example, if a robot arm is moving, there needs to be a perception mechanism in place that continuously provides a dynamic obstacle map of the surrounding environment in order to prevent the robot arm colliding with it. It doesn't matter if the system cannot provide a fine triangular mesh every couple of milliseconds because the surface reconstruction algorithms cannot respect these computational constraints, but instead, a rougher representation using 3D grids (voxels), or bounding spheres needs to be created and provided. Though not as precise, it will keep the robot safe without crashing and thus prevent a catastrophe. Additionally, different tasks require different levels of detail. While navigating to the closest table needs only the rough position of the table, manipulating an object located on it needs to model finer details with regards to the exact table bounds, table height, as well as the object that is to be grasped.

These are just a few of the theoretical requirements of the system architecture for a Semantic 3D Object Mapping kernel. What is obvious though, is that if the design of the architecture adheres to these standards, it will be possible to construct particular mapping systems that tackle and solve different applications in an efficient manner. Simply stated, there is no single architecture that can be applied to all problems.

From the list of requirements presented above, we can formulate the definition of a *mapping kernel*. Since we cannot expect to have algorithms that perform perfectly in all possible conditions, we would like to create a *pool of methods* that work very well for a particular situation instead, and use them in conjunction with each other whenever more complex problems are to be solved. Because we already tackled the modularity issue above, we expect methods belonging to the same class to be swappable between each other and provide information through a series of standard, well defined interfaces. For example, though there might be two different ways to segment an object of interest from the planar surface it resides on, both methods should be instantiated and their results interpreted in the same way, so that a higher level executive that controls the perception system can use one or the other without any changes to the code. This collection of algorithms together with their respective interfaces forms what we call the Semantic 3D Object Mapping kernel used for the remaining of this book. Based on the application requirements, we will create and instantiate processing pipelines of such algorithms, which will lead to unique architectures for every particular problem that we need to solve. This does not imply that the proposed framework is not well defined, but rather that we prefer flexible and efficient solutions for a given problem.

From the data acquisition point of view, an important aspect that has to be addressed at the task executive level that is controlling the sensors and the mobile platform, concerns the amount of overlap that individual data frames must have with respect to each other. Because data acquisition routines are inherently 2D or 2.5D, it is impossible to obtain a complete 360° view representation for all the objects present in the world. Therefore, the system must assure that the individually acquired frames are overlapping with a large degree, in order to ensure that data registration algorithms will succeed in finding a correct solution every time.

An example of a complex system architecture that drives a good part of the work presented in this book is shown in Figure 3.1. Though applicable to data coming from indoor environments (in particular kitchens) and to the problem of creating a semantic functional representation of the world useful for mobile manipulation and grasping applications, most of the concepts presented here have been shown to give good results for other problems as well [RMCB09, RBM$^+$09, RSG$^+$09, MRJ$^+$09], including outdoor datasets [RBMB08, MRB09, BRBB09, RMBB08a].

As seen from Figure 3.1, the input point cloud data is provided to the mapping system via a point cloud acquisition module. The point cloud can be either generated by the sensing device itself, or in the case where 2D laser sensors are used, converted into a consistent Euclidean representation using specific software tools. The input data is then filtered using different methods to remove spurious measurements and outliers, such as jump edges, in a sparse outlier removal module. Based on the goals of the application, the data can then be resampled (downsampled usually) so that the complexity of the problem is reduced, followed by a set of feature estimation steps, including: normal and curvature estimation and consistent orientation, a feature persistence analysis, and the estimation of complex features such as Point Feature Histogram representations (see Chapter 4).

The next step in the proposed architecture consists in the registration of individual interpreted point cloud data views into an integrated and complete point cloud model of the world, correcting the imprecise alignment caused by odometry. After registration, the complete model could be resampled again to account for registration errors and point density variations.

In general, the above mentioned steps can be estimated for any input dataset, independent of the problem that needs to be solved, and thus we call the collection of such steps as being part of a Geometric Mapping framework. In contrast, the remaining methods are instantiated based on various needs and particular application goals, as they tend to extract specific "functional" information from the environment data.

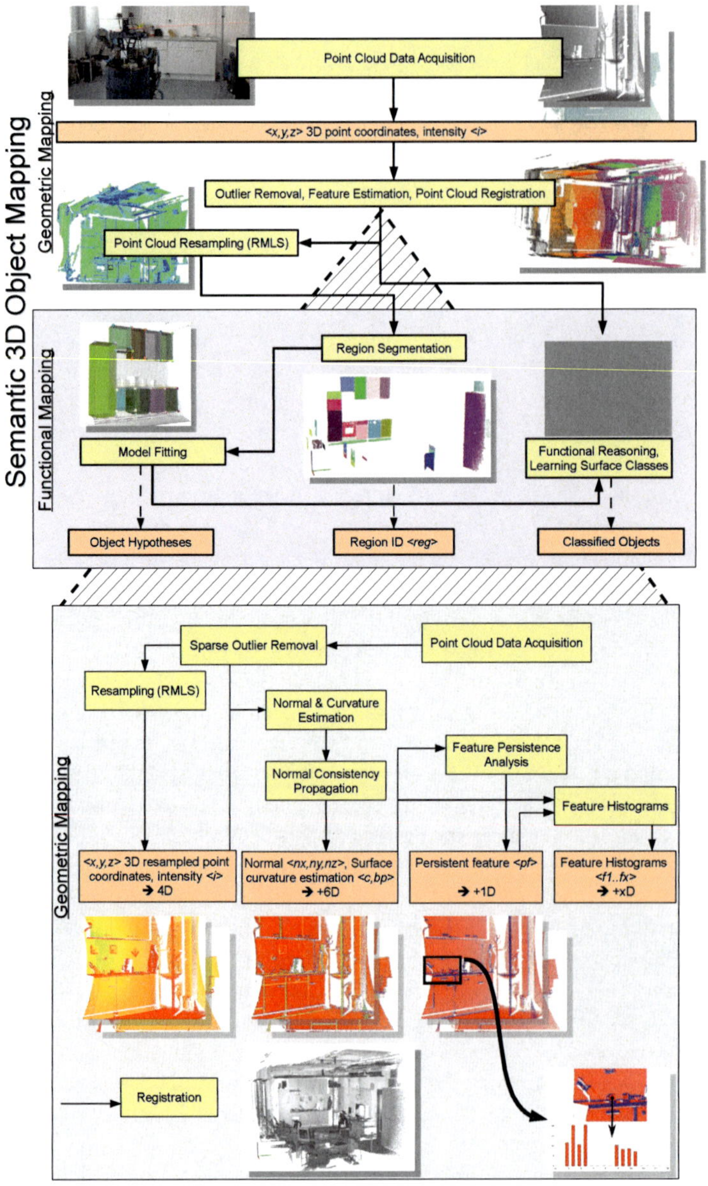

Fig. 3.1 An example of a complex system architecture for a Semantic 3D Object Mapping system for the particular problem of mapping in indoor environments (in particular kitchens) as presented in this book

As part of the Functional Mapping framework, the system can generate object hypotheses and include them as individual entities in the map. In addition, rectangular planes and boxes of specific size ranges can be assessed as candidates for specific parts of the world, such as cupboards and tables. Their actual class designation can then be given either through a set of heuristic common sense rules, or by learning models from sets of training data and using them to classify new data instances. To compute this semantic object environment model, the robot monitors acquired point cloud data views over time, and integrates them into the existing map.

All the above mentioned steps will be explained in detail in the next chapters.

Chapter 4
3D Point Feature Representations

> *"One new feature or fresh take can change everything."*
>
> NEIL YOUNG

In their native representation, *points* as defined in the concept of 3D mapping systems are simply represented using their Cartesian coordinates x, y, z, with respect to a given origin. Assuming that the origin of the coordinate system does not change over time, there could be two points $\boldsymbol{p}_1$ and $\boldsymbol{p}_2$, acquired at t_1 and t_2, having the same coordinates. Comparing these points however is an ill-posed problem, because even though they are equal with respect to some distance measure (*e.g.* Euclidean metric), they could be sampled on completely different surfaces, and thus represent totally different information when taken together with the other surrounding points in their vicinity. That is because there are no guarantees that the world has not changed between t_1 and t_2. Some acquisition devices might provide extra information for a sampled point, such as an intensity or surface remission value, or even a color, however that does not solve the problem completely and the comparison remains ambiguous.

Applications which need to compare points for various reasons require better characteristics and metrics to be able to distinguish between geometric surfaces. The concept of a 3D point as a singular entity with Cartesian coordinates therefore disappears, and a new concept, that of *local descriptor*, takes its place. The literature is abundant of different naming schemes describing the same conceptualization, such as *shape descriptors*, or *geometric features*, but for the remaining of this book they will be referred to as *point feature representations*.

The theoretical formulation of a feature representation for a given point $\boldsymbol{p}_q$ is given as follows. Let $\boldsymbol{p}_q$ be the query point, and $\mathcal{P}^k = \{\boldsymbol{p}_1^k \cdots \boldsymbol{p}_2^k\}$ a set of points located in the neighboring vicinity of $\boldsymbol{p}_q$. The concept of a neighbor is given as:

$$\left\|\boldsymbol{p}_i^k - \boldsymbol{p}_q\right\|_x \leq d_m \tag{4.1}$$

where d_m is a specified maximum allowed distance from the neighbor to the query point, and $\|\cdot\|_x$ is an example L_x Minkowski norm (though different other distance norms can be used). Additionally, the number of neighbors in $\mathcal{P}^k$ can be capped

R.B. Rusu: *Semantic 3D Object Maps for Everyday Robot Manipulation*, STAR 85, pp. 33–60.
DOI: 10.1007/978-3-642-35479-3_4

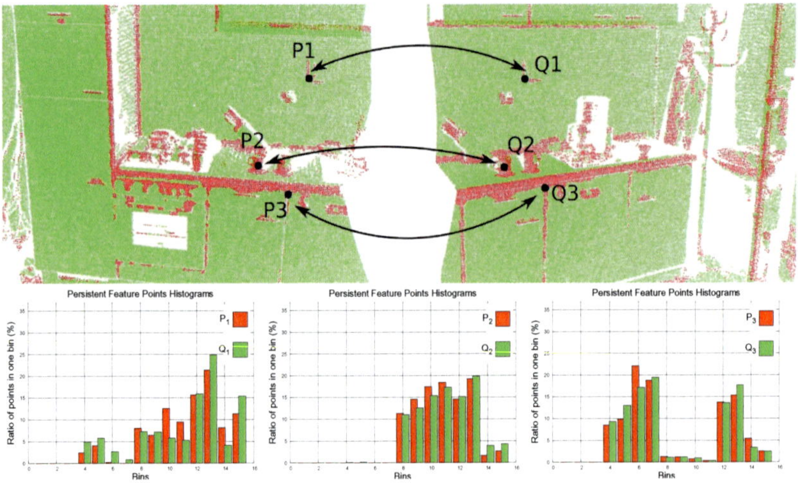

Fig. 4.1 Point feature representations for three pairs of corresponding points $(\boldsymbol{p}_i, \boldsymbol{q}_i)$ on 2 different point cloud datasets

to a given value k. A point feature representation can then be described as a vector function F, describing the local geometric information captured by $\mathcal{P}^k$, around $\boldsymbol{p}_q$:

$$F(\boldsymbol{p}_q, \mathcal{P}^k) = \{x_1, x_2, x_3 \cdots x_n\} \tag{4.2}$$

where x_i, $i \in \{1 \cdots n\}$ represents the dimension i of the resultant feature vector representation. Comparing two different points $\boldsymbol{p}_1$ and $\boldsymbol{p}_2$ then results in comparing the difference in some space between their feature vectors F_1 and F_2. Let Γ be the similarity measure describing the difference between $\boldsymbol{p}_1$ and $\boldsymbol{p}_2$, and d a distance metric, then:

$$\Gamma = d(F_1, F_2) \tag{4.3}$$

As d tends to some minimum, say $d \to 0$, the two points will be considered as being similar with respect to their feature representations. If d is large, then the points are to be considered distinct from each other – *i.e.*, they represent different surface geometries.

By including the surrounding neighbors, the underlying sampled surface geometry can be inferred and captured in the feature formulation, which contributes to solving the ambiguity comparison problem. Ideally, the resultant features would be very similar (with respect to some metric) for points residing on the same or similar surfaces, and different for points found on different surfaces, as shown in Figure 4.1.

A "good" point feature representation distinguishes itself from a "bad" one, by being able to capture the same local surface characteristics in the presence of:

- **rigid transformations** – that is, 3D rotations and 3D translations in the data should not influence the resultant feature vector F estimation;
- **varying sampling density** – in principle, a local surface patch sampled more or less densely should have the same feature vector signature;
- **noise** – the point feature representation must retain the same or very similar values in its feature vector in the presence of mild noise in the data.

The next sections address the problem of feature estimation, and explain the process required in obtaining the set of neighbors $\mathcal{P}^k$ for a given query point $\boldsymbol{p}_q$.

4.1 The "Neighborhood" Concept

Conceptually, the problem of determining the point's neighbors $\mathcal{P}^k$ is closely related to the specific metric space that needs to be used. In general, the definition given in Equation 4.1 stands, with the Euclidean L_2 norm being the most popular instantiation. This means that in order to determine the k closest points of a query point $\boldsymbol{p}_q \in \mathcal{P}$, all distances from $\boldsymbol{p}_q$ to all the other points in $\mathcal{P}$ must be estimated and ordered, with the first smallest k ones corresponding to the closest set of points $\mathcal{P}^k$.

This brute-force process however is extremely costly and it makes little sense to use it for applications where closest point queries are frequent. Additionally, because of the small errors present in the acquired data, the true nearest neighbor of a point might be different than the one that has the smallest distance, especially if the input data is dense enough. This promotes the use of *approximate nearest neighbor* solutions, and in particular, orthogonal decomposition methods such as the ones presented in [AM93, AMN$^+$98, ML09]. These approaches employ a parameter $\epsilon \geq 0$ that describes the acceptable error bound for the solution to be found. For any returned neighboring point $\boldsymbol{p}_i^k$, the ratio between the distance to it and the true nearest neighbor $\boldsymbol{q}^k$ is at most $1+\epsilon$, that is:

$$\|\boldsymbol{p}_i^k - \boldsymbol{p}_q\|_x \leq (1+\epsilon)\|\boldsymbol{q}^k - \boldsymbol{p}_q\|_x \tag{4.4}$$

There are many solutions that can speed up the computation even further and return extremely fast results using these error margins. From an implementation point of view, the estimated distances are usually squared such that slower square root calculus is avoided, values can be cached as proposed in [And07] to speed up subsequent searches, or multiple randomized trees can be created at once to help split the dimensions on which the data has the greatest variance [ML09, SAH08].

From an usage point of view, there are two specific application examples for the determination of $\mathcal{P}^k$ for a query point $\boldsymbol{p}_q$, namely:

- determine the closest k neighbors of the query point (k search);
- determine all the k neighbors of the query point up to a radius r from it (radius search).

The latter is of particular interest for 3D feature estimation, because it attempts to capture the data on the same surface patch, independently on the number of points sampled, distance, or angle with respect to the sensor. An example is shown in Figure 4.2 where for a given query point $\boldsymbol{p}_q$ (marked with red), the set of $\mathcal{P}^k$ neighbors in a specific given radius r is selected for the purpose of estimating the normal to the surface sampled around $\boldsymbol{p}_q$ (see Section 4.3 for the mathematical details).

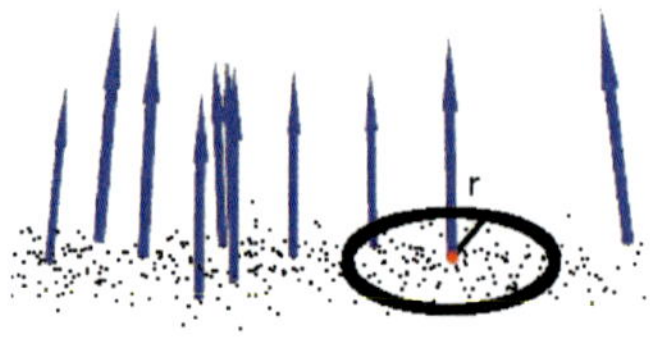

Fig. 4.2 Example of a radius r search to determine the set of neighbors $\mathcal{P}^k$ for the query point (marked with red). $\mathcal{P}^k$ is used to estimate an approximative surface normal at the query point (blue arrow).

The specifics of the nearest-neighbor estimation problem raise the question of the *right scale* factor: given a sampled point cloud dataset $\mathcal{P}$, what are the correct k or r values that should be used in determining the set of nearest neighbors $\mathcal{P}^k$ of a point? This issue is of extreme importance and constitutes a limiting factor in the automatic estimation (*i.e.*, without user given thresholds) of a point feature representation. To better illustrate this issue, Figures 4.3 and 4.4 present the effects of selecting a smaller scale (*i.e.*, small r or k) versus a larger scale (*i.e.*, large r or k). The left part of the figures depicts a reasonable well chosen scale factor, with estimated surface normals approximatively perpendicular for the two planar surfaces (Figure 4.3), and small edges visible all across the table (Figure 4.4). If the scale factor however is too big (right part), and thus the set of neighbors $\mathcal{P}^k$ is larger covering points from adjacent surfaces, the estimated point feature representations get distorted, with rotated surface normals at the edges of the two planar surfaces (Figure 4.3), and smeared edges and suppressed fine details (Figure 4.4).

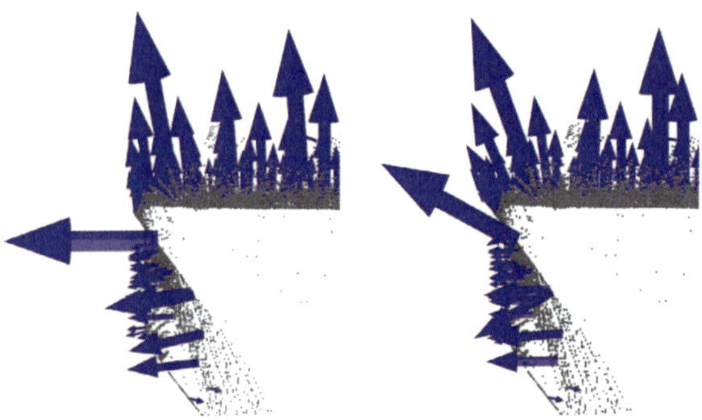

Fig. 4.3 Example of estimated surface normals for a subset of points in a given $\mathcal{P}$ using: a) a small (good) scale factor (left); b) a large (bad) scale factor (right)

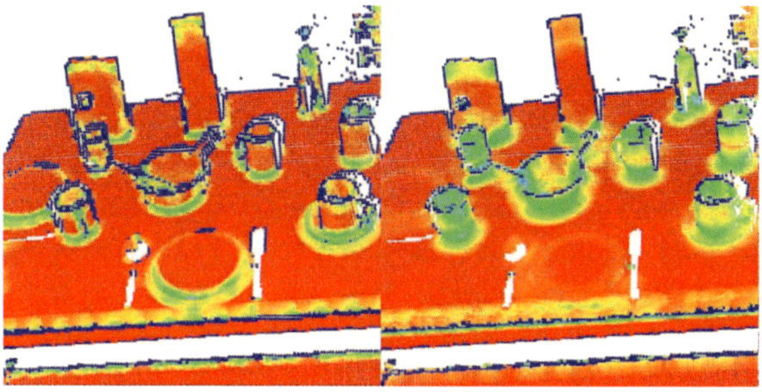

Fig. 4.4 Example of estimated surface curvatures for a subset of points in a given $\mathcal{P}$ using: a) a small (good) scale factor (left); b) a large (bad) scale factor (right). See Section 4.3 for the mathematical details.

Some of these particularities have been investigated in initiatives such as [LUVH05] or [MN03] in the context of surface normal estimation, where the authors attempt to automatically estimate the correct scale by iteratively determining the patch density for increasing radii r. These methods are slow however, and still rely on assumptions with regards to a somewhat constant point density throughout the datasets. Moreover, they have been applied only to surface normal estimation which is just one of the many feature representations. A comprehensive analysis on multi-scale statistical approaches is also given in [Unn08]. Section 4.6 will present a more in-depth discussion regarding the above points, but for now it suffices to assume that the scale for the determination of a point's neighborhood has to be selected based on the level of detail required by the application. Simply put, if the curvature at the edge between the handle of a mug and the cylindrical part is important, the scale factor needs to be small enough to capture those details, and large otherwise.

4.2 Filtering Outliers

Before attempting to estimate the characteristics of a point with respects to its surrounding neighbors, it is important to analyze if the surrounding neighbors are good representations of the underlying sampled surface. This analysis can range from simple thresholding values that do not respect certain constraints, to performing a more detailed mathematical analysis on the uncertainties in the measured position like in [BBL09]. In the interest of generality, the methods given here expect minimal assumptions about the underlying acquisition device, and can thus be in general applied to any point cloud dataset $\mathcal{P}$.

The majority of point feature representations require an absolute minimum of $k \geq k_{min}$ neighbors in the vicinity of a query point $\boldsymbol{p}_q$ in order to be computable. Due to the varying data density present in 2.5D scans however, there will be several points which cannot respect these constraints. These either reside on surfaces with high reflective properties such as shiny metals, or appear as outliers at the transition between two surfaces due to occlusions. The latter are also known as jump edges, depth discontinuities, or occlusion boundaries in the literature. Removing these points from the point cloud leads to an overall faster processing time, due to the reduced remaining number of points.

The literature is abundant in methods for the removal or marking of jump edges, either based on segmentation principles [PRLLM08, Pet02], or by exploiting the properties of a particular sensor [MDH$^+$08], with an interesting overall comparison given in [THA07], to name a few. From the feature estimation point of view however, a unified formulation able to remove all points in sparse density areas, including jump edges, would solve the problem better.

The proposed solution is based on a statistical analysis of each point's neighborhood $\mathcal{P}^k$ as we previously proposed in [RMB$^+$08a]. For each point $\boldsymbol{p}_q \in \mathcal{P}$, the mean distance $\bar{d}$ to its k closest neighbors is first computed. Then, a distribution over the mean distance space for the entire point cloud $\mathcal{P}$ is assembled and its mean μ_k and standard deviation σ_k are estimated. The goal is to retain the points whose mean distance $\bar{d}$ to the closest k neighbors is similar to the one for the rest of the points. As this describes a measure of the underlying point cloud density surrounding a point, the remaining point cloud $\mathcal{P}^*$ is simply estimated as follows:

$$\mathcal{P}^* = \{\boldsymbol{p}_q^* \in \mathcal{P} | \; (\mu_k - \alpha \cdot \sigma_k) \leq \bar{d}^* \leq (\mu_k + \alpha \cdot \sigma_k)\} \tag{4.5}$$

where α is a desired density restrictiveness factor. An illustrative example for the effects of the statistical analysis and filtering is shown in Figure 4.5. The left part of the figure presents a point cloud dataset $\mathcal{P}$ containing several areas with lower point densities (marked with circles) encapsulating either jump edges or spurious measurements. The middle part of the figure presents the resultant point cloud $\mathcal{P}^*$, after the removal of points residing in sparse density areas. The value of α was set to 1 for the purpose of these experiments.

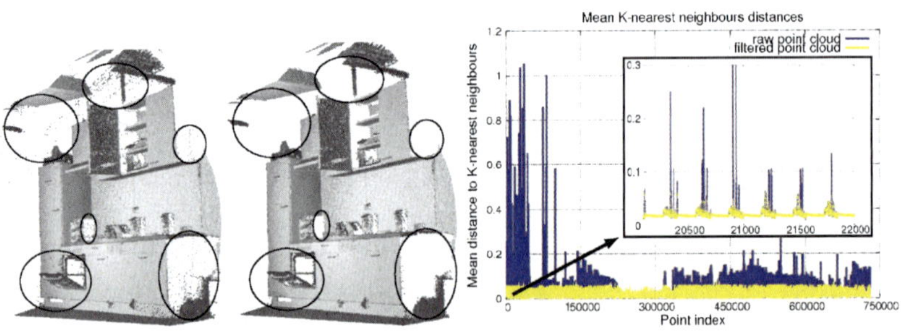

Fig. 4.5 A point cloud dataset $\mathcal{P}$ containing many outliers (left). The resultant filtered dataset $\mathcal{P}^*$ after statistical analysis and filtering (middle), with number of points left: 728070 out of 730664, *i.e.* 99.64%. The mean distances to $k = 30$ neighbors before and after removal using $\alpha = 1$. Remission values are shown in grayscale in both scans.

An alternative method for removing bad data, this time in the context of gross erroneous measurements (*i.e.*, resultant points where there is no surface), is to consider additional scans taken from the same or a very close position, and infer the errors by fusing the information between the scans. This is only possible of course, if the individually acquired point clouds $\mathcal{P}_i$ can be transformed so that they lie in the same coordinate system.

Assuming that the world did not change between the first scan $\mathcal{P}_i$ and the subsequent $\mathcal{P}_{i+1}$, and that their acquisition viewpoints are identical, the simplest form of filtering outliers could be employed as follows. For each point $\boldsymbol{p} \in \mathcal{P}_i$, search for its closest correspondent neighbor $\boldsymbol{q} \in \mathcal{P}_{i+1}$. If the distance between $\boldsymbol{p}$ and $\boldsymbol{q}$ is larger than a user specified threshold, then $\boldsymbol{p}$ is an outlier and can be safely removed from $\mathcal{P}_i$. Additionally, all the remaining points could be averaged between the two scans.

For non-identical acquisition points, a viable solution is to apply a raycasting based scheme such as the one we previously proposed in [BRBB09]. Given a point cloud dataset $\mathcal{P}_i$, an octree representation O_i is created, such that all points $\boldsymbol{p}_i \in \mathcal{P}_i$ are associated with a voxel $v_i \in O_i$ (see Figure 4.6).

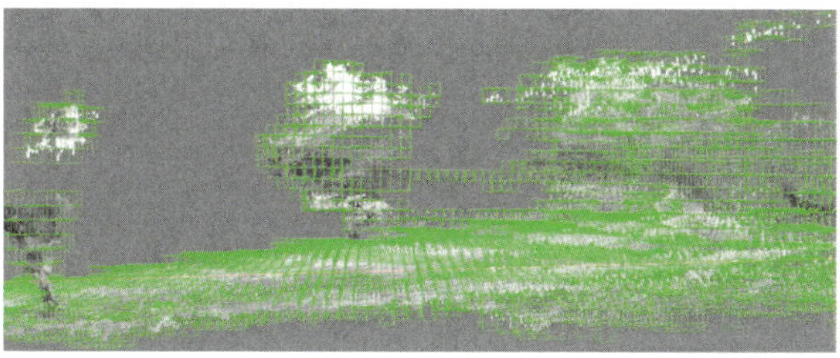

Fig. 4.6 Voxelizing a point cloud dataset for the purpose of raycasting-based filtering. The resultant octree is shown in green.

Every new dataset $\mathcal{P}_{i+1}$ is first transformed into the same coordinate system as $\mathcal{P}_i$ (see Chapter 5), followed by the creation of a new O_{i+1} octree representation. Given that the two acquisition viewpoints for $\mathcal{P}_i$ and $\mathcal{P}_{i+1}$, namely v_{p_i} and $\mathsf{v}_{p_{i+1}}$ are now in the same coordinate system as well, a ray (*i.e.*, line) from $\mathsf{v}_{p_{i+1}}$ to each of the voxels $v_{i+1} \in O_{i+1}$ is created. Due to the volumetric octree representation, the line will intersect in its path: a) free empty voxels of either O_i or O_{i+1}, or b) occupied voxels $v_i^* \in O_i$. By making the same assumptions about the world being unchanged between the time the two point cloud datasets were acquired, all intersected voxels $v_i^* \in O_i$ can be uniformly assumed to be outliers and the points $\boldsymbol{p}_i$ located in them can be removed. Figure 4.7 presents the raycasting process for two subsequently acquired datasets as described above. The points discovered to be outliers are marked with red.

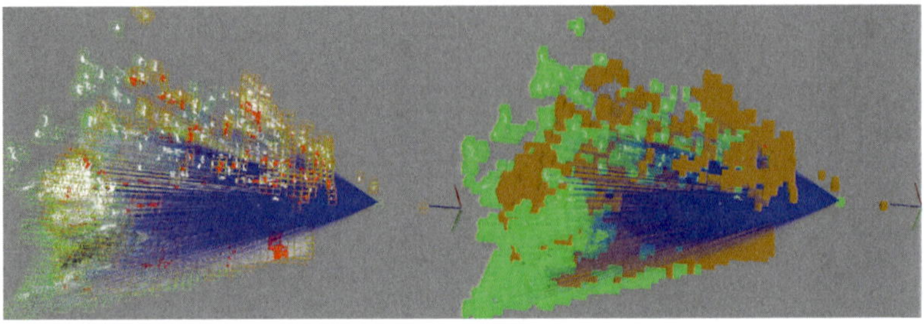

Fig. 4.7 Two subsequently acquired datasets ($\mathcal{P}_i$ and $\mathcal{P}_{i+1}$) and their corresponding octrees (O_i and O_{i+1}) shown in brown (i) and green respectively ($i+1$). The blue lines represent the rays casted from the acquisition viewpoint $\mathrm{v}_{p_{i+1}}$ to the voxels $v_{i+1} \in O_{i+1}$. The filtered points (outliers) are shown in red. The left part of the figure presents the wireframe interpretation of the scene in the right.

4.3 Surface Normals and Curvature Estimates

Once determined, the neighboring points $\mathcal{P}^k$ of a query point $\boldsymbol{p}_q$ can be used to estimate a local feature representation that captures the geometry of the underlying sampled surface around $\boldsymbol{p}_q$. An important problem in describing the geometry of the surface is to first infer its orientation in a coordinate system, that is, estimate its normal. Surface normals are important properties of a surface and are heavily used in many areas such as computer graphics applications to apply the correct light sources that generate shadings and other visual effects.

Though many different normal estimation methods exist (see [KAWB09] for a recent comparison for normal estimation in 3D range data), the simplest method is based on the first order 3D plane fitting as proposed by [BC94]. The problem of determining the normal to a point on the surface is approximated by the problem of estimating the normal of a plane tangent to the surface, which in turn becomes a least-square plane fitting estimation problem in $\mathcal{P}^k$ [Sha98]. The plane is represented as a point $\boldsymbol{x}$ and a normal vector $\vec{\boldsymbol{n}}$, and the distance from a point $\boldsymbol{p}_i \in \mathcal{P}^k$ to the plane is defined as $d_i = (\boldsymbol{p}_i - \boldsymbol{x}) \cdot \vec{\boldsymbol{n}}$. The values of $\boldsymbol{x}$ and $\vec{\boldsymbol{n}}$ are computed in a least-square sense so that $d_i = 0$. By taking:

$$\boldsymbol{x} = \overline{\boldsymbol{p}} = \frac{1}{k} \cdot \sum_{i=1}^{k} \boldsymbol{p}_i \tag{4.6}$$

as the centroid of $\mathcal{P}^k$, the solution for $\vec{\boldsymbol{n}}$ is given by analyzing the eigenvalues and eigenvectors of the covariance matrix $\mathcal{C} \in \mathbb{R}^{3\times 3}$ of $\mathcal{P}^k$, expressed as:

$$\mathcal{C} = \frac{1}{k} \sum_{i=1}^{k} \xi_i \cdot (\boldsymbol{p}_i - \overline{\boldsymbol{p}}) \cdot (\boldsymbol{p}_i - \overline{\boldsymbol{p}})^T,\ \mathcal{C} \cdot \vec{\mathrm{v}}_j = \lambda_j \cdot \vec{\mathrm{v}}_j,\ j \in \{0,1,2\} \tag{4.7}$$

The term ξ_i represents a possible weight for p_i, and usually equals 1. $\mathcal{C}$ is symmetric and positive semi-definite, and its eigenvalues are real numbers $\lambda_j \in \mathbb{R}$.

The eigenvectors $\vec{v_j}$ form an orthogonal frame, corresponding to the principal components of $\mathcal{P}^k$. If $0 \leq \lambda_0 \leq \lambda_1 \leq \lambda_2$, the eigenvector $\vec{v_0}$ corresponding to the smallest eigenvalue λ_0 is therefore the approximation of $+\vec{\boldsymbol{n}} = \{n_x, n_y, n_z\}$ or $-\vec{\boldsymbol{n}}$. Alternatively, $\vec{\boldsymbol{n}}$ can be represented as a pair of angles (ϕ, θ) in spherical coordinates [HLLS01], as:

$$\phi = \arctan\left(\frac{n_z}{n_y}\right), \theta = \arctan\frac{\sqrt{(n_y^2 + n_z^2)}}{n_x} \tag{4.8}$$

In general, because there is no mathematical way to solve for the sign of $\vec{\boldsymbol{n}}$, the orientation of the normal computed via Principal Component Analysis (PCA) as shown above is ambiguous, and not consistently oriented over a point cloud dataset $\mathcal{P}$. Figure 4.8 presents these effects on two sections of a larger dataset representing a part of a kitchen environment. The right part of the figure presents the Extended Gaussian Image (EGI), also known as the normal sphere, which describes the orientation of all normals $\vec{\boldsymbol{n}}_i$ from $\mathcal{P}$. Since the datasets are 2.5D and have thus been acquired from a single viewpoint v_p, normals should be present only on half of the sphere in the EGI. However, due to the orientation inconsistency, they are spread across the entire sphere.

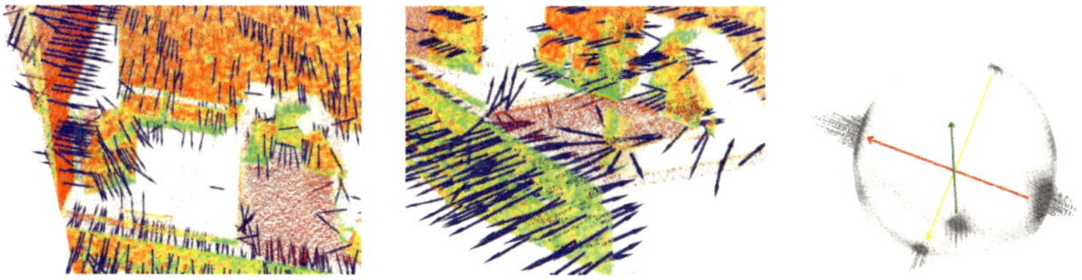

Fig. 4.8 Left and middle: inconsistently oriented surface normals $\vec{\boldsymbol{n}}_i$ acquired using PCA in a 2.5D dataset $\mathcal{P}$. Right: the resultant EGI (normal sphere) for $\mathcal{P}$.

The solution to this problem is trivial if the viewpoint v_p is in fact known. To orient all normals $\vec{\boldsymbol{n}}_i$ consistently towards the viewpoint, they need to satisfy the equation:

$$\vec{\boldsymbol{n}}_i \cdot (\mathsf{v}_p - \boldsymbol{p}_i) > 0 \tag{4.9}$$

Figure 4.9 presents the results after all normals in the datasets from Figure 4.8 have been consistently oriented towards the viewpoint. In situations where the viewpoint information is not available, the problem is a bit more difficult to solve. One possible solution is given by [HDD$^+$92], and consists in modeling the consistency as a graph optimization problem. The key idea is to consider that two data points $\boldsymbol{p}_i$ and $\boldsymbol{p}_j$ belonging to a smooth surface and geometrically close, need to have their normals consistently oriented, that is:

$$\vec{\boldsymbol{n}}_i \cdot \vec{\boldsymbol{n}}_j \approx 1 \tag{4.10}$$

Fig. 4.9 Left and middle: consistently re-oriented surface normals $\vec{n}_i$ satisfying equation 4.9. Right: the resultant EGI (normal sphere) after re-orientation for $\mathcal{P}$.

In general, this assumption holds true for densely sampled datasets representing smooth surfaces. Each point $p_i \in \mathcal{P}$ is therefore modeled as a node in a graph, with edge costs being set to the Euclidean distances between the points. This graph formulation that tends to connect neighbors is found to be the Euclidean Minimum Spanning Tree (EMST). Starting from a random node encapsulating p_i that is guaranteed to have a correct normal orientation $\vec{n}_i$, the orientation of its adjacent nodes is propagated by traversing the minimal spanning tree of the graph and changing the sign of each normal which satisfies the equation:

$$\vec{n}_i \cdot \vec{n}_j < 0 \Rightarrow \vec{n}_j = -\vec{n}_j \tag{4.11}$$

As the authors note [HDD+92], setting the cost of graph edges to Euclidean distances can sometimes lead to failures in the propagation orientation. A better edge cost e_g is therefore formulated as:

$$e_g = 1 - |\vec{n}_i \cdot \vec{n}_j| \tag{4.12}$$

The proposed cost is non-negative and has the property that its value is small for unoriented nearly parallel normals, therefore enabling a better traversal of the minimal spanning tree for a consistent normal propagation. Figure 4.10 presents the Riemannian graphs created with the method proposed in [HDD+92] for the datasets presented in Figure 4.8. Due to the fact that the surfaces sampled are non-smooth, the resultant graph is not perfect. Moreover, the algorithm has a big computational complexity for large datasets like these.

In addition to surface normal estimates, the eigenanalysis (PCA) presented earlier can also be used to infer different measures of the surface curvature around p_q. In general there are many ways to define the curvature of a surface at a specific point, but they usually require the surface to be already represented as a triangular mesh, and not just by points sampled on it. The work of [DHKL01] defines the Gaussian K and mean $\overline{H}$ curvature of a smooth surface at a certain vertex point as:

$$K = \kappa_1 \kappa_2,\ \overline{H} = \frac{\kappa_1 + \kappa_2}{2},\ \overline{H}^2 \geq K \tag{4.13}$$

where κ_1 and κ_2 are the principal curvatures of the surface. The pair $\{K, \overline{H}\}$ however provides a poor representation since the estimated values are strongly

Fig. 4.10 Riemannian graphs for general sampled surfaces used for the purpose of consistent normal orientation

correlated. A different curvature formulation [KvD92] is given in the form of the shape index $S_I \in [0, 1]$:

$$S_I = \frac{1}{2} - \frac{1}{\pi}\arctan\left(\frac{\kappa_1 + \kappa_2}{\kappa_1 - \kappa_2}\right), \kappa_1 \geq \kappa_2 \tag{4.14}$$

Unfortunately, the above expressions are sensitive to noise [HLLS01] and cannot be estimated from a set of sampled points $\mathcal{P}^k$ directly. A solution is to use the eigenvalues λ_j of the covariance matrix C as approximations of the surface variation around $\boldsymbol{p}$. If $\lambda_0 = \min(\lambda_j)$, the variation of a point $\boldsymbol{p}$ along the surface normal $\vec{\boldsymbol{n}}$ can be estimated as:

$$\sigma_p = \frac{\lambda_0}{\lambda_0 + \lambda_1 + \lambda_2} \tag{4.15}$$

This ratio between the minimum eigenvalue and the sum of the eigenvalues approximates the change of curvature in a neighborhood $\mathcal{P}^k$ centered around $\boldsymbol{p}$ [PGK02], and is invariant under rescaling. Small values of σ_p indicate that all points in $\mathcal{P}^k$ are on the plane tangent to the surface.

The estimated σ_{p_i} values are dependent on the choice of k and the estimation of C, and can be highly influenced by noise. An alternative scheme that we proposed in [RMB$^+$08a] is to attempt to suppress noisy data by adapting the covariance matrix C based on the inlier point distances to a best fit plane. This means that instead of estimating the tangent least-squares plane to the surface, the algorithm attempts to find the best plane model fit in a sample consensus framework, given a maximum distance deviation d_{max} from each point $\boldsymbol{p}_i$ to the plane. Once the model has been found, the points in $\mathcal{P}^k$ can be classified as inliers if their distance to the plane $d_i \leq d_{max}$, and outliers otherwise. To influence the computation of the covariance matrix C, the weight term ξ_i can be rewritten as follows:

$$\xi_i = \begin{cases} \exp(-\frac{d_i^2}{\mu^2}), \boldsymbol{p}_i^k \text{ outlier} \\ 1, \boldsymbol{p}_i^k \text{ inlier} \end{cases} \tag{4.16}$$

where μ is the mean distance from the query point $\boldsymbol{p}_q$ to all its neighbors $\boldsymbol{p}_i^k \in \mathcal{P}^k$, and d_i is the distance from $\boldsymbol{p}$ to its $\boldsymbol{p}_i^k$ neighbor.

Our method has the advantage that it takes into account the distance from every outlier (found using the robust estimator above) to the query point, thus changing the model slightly. This will guarantee that correct normals are estimated even in the proximity of sharp edges. The search for planar models is performed using a RMSAC (Randomized M-Estimator SAmple Consensus) estimation method, which adds a $T_{d,d}$ test as proposed in [CM02] to an MSAC estimator [TZ00]. Figure 4.11 presents the effect of the k-neighborhood support selection in a noisy point cloud for surface curvature estimation, using standard techniques (left), and using our proposed method (right). Because in our method the surface curvature is estimated based on a better k neighborhood support, edges are much sharper. This has a big influence on the quality of point cloud segmentation in different regions of interest, as we will see in Chapter 6.

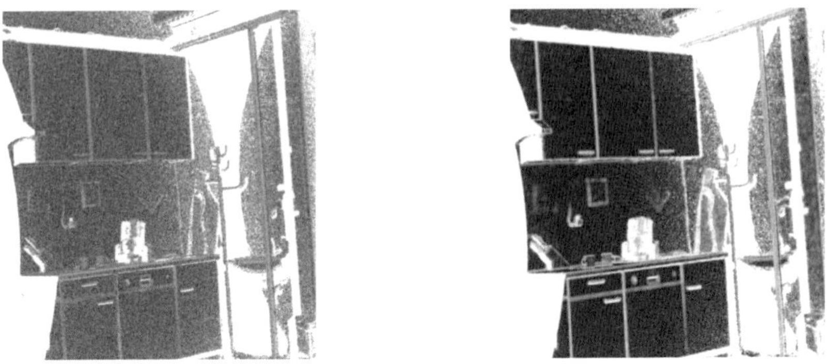

Fig. 4.11 Estimating surface curvatures for a noisy point cloud: using all $k = 30$ neighbors as inliers (left), and selecting automatically the best k-neighborhood support using our method (right). In both scans, curvature values are shown in grayscale: low curvatures are darker.

4.4 Point Feature Histograms (PFH)

As point feature representations go, surface normals and curvature estimates are somewhat basic in their representations of the geometry around a specific point. Though extremely fast and easy to compute, they cannot capture too much detail, as they approximate the geometry of $\mathcal{P}^k$ with only a few values. As a direct consequence, most scenes will contain many points with the same or very similar feature values, thus reducing their informative characteristics. Even if the feature estimation would be able to cope with noisy datasets, it can still be easily deduced that applications who rely on these minimal dimensional features will deal with multiple and false correspondences and will be prone to failure.

To alleviate the above, several communities have proposed a multitude of different types of point feature representations that would allow a better representation of the neighborhood surface. Section 4.7 attempts to address some of these related initiatives and to explain the differences between them and our proposed approach. In general though, it would be ideal to have an informative label as a feature representation for a point, such as: point lying on an edge, point lying on a sphere or a plane, and so on. Using such feature representations and sets of geometric constraints between them, the probability of uniquely defining surface geometries would increase. Therefore one of the goals of our work is to identify a multi-dimensional feature space with a high discriminative power, such that point feature representations for data points sampled on the same surface can be grouped in the same class, while data points on different surfaces must be assigned to different classes. The representation proposed in this book is inspired by the global descriptor work on synthetic object recognition presented in [WHH03].

To formulate the new feature space, the concept of a *dual-ring neighborhood* is first introduced. Let $\mathcal{P}$ be a set of 3D points with $\{x_i, y_i, z_i\}$ geometric coordinates. A point $\boldsymbol{p}_i \in \mathcal{P}$ is said to have a *dual-ring neighborhood* if:

$$(\exists) r_1, r_2 \in \mathbb{R},\ r_1 < r_2, \text{ such that } \begin{cases} r_1 \Rightarrow \mathcal{P}^{k_1} \\ r_2 \Rightarrow \mathcal{P}^{k_2} \end{cases}, \text{with } 0 < k_1 < k_2 \tag{4.17}$$

The two radii r_1 and r_2 are used to determine two distinct layers of feature representations for $\boldsymbol{p}_i$. The first layer has already been described in the previous section and represents the surface normal at the query point, obtained from the Principal Component Analysis of the neighborhood patch $\mathcal{P}^{k_1}$. The second layer comprises the Point Feature Histogram (PFH) representation, explained in this section.

The goal of the PFH formulation is to encode the $\mathcal{P}^{k_2}$ neighborhood's geometrical properties by generalizing the mean curvature around $\boldsymbol{p}_i$ using a multi-dimensional histogram of values. This highly dimensional hyperspace provides an informative signature for the feature representation, is invariant to the 6D pose of the underlying surface, and copes very well with different sampling densities or noise levels present in the neighborhood.

A Point Feature Histogram representation is based on the relationships between the points in $\mathcal{P}^{k_2}$ and their normals. Simply put, it attempts to capture as best as possible the sampled surface variations by taking into account all the interactions between the directions of the estimated normals. The resultant hyperspace is thus dependent on the quality of the surface normal estimations at each point. The idea is somewhat similar to the Support Vector Machine theory presented in Appendix C.1, where the input data is first transformed into a higher order dimensional space through the usage of special functions called kernels.

In general, there could be many ways to capture the variations of surface normals in a local patch. Since the purpose of our formulation is not to average values out, like other similar representations (see Section 4.7), but to model the intrinsic properties of the surface in detail, a measure of variation will be estimated for each pair of points $\boldsymbol{p}_i$ and $\boldsymbol{p}_j$ from $\mathcal{P}^{k_2}$. The first step in estimating the PFH for a point $\boldsymbol{p}_i$ is to therefore estimate all surface normals $\vec{\boldsymbol{n}}_i$ from the points in $\mathcal{P}^{k_2}$. This can be computed either on demand, or done as a preliminary step for each $\boldsymbol{p}_i \in \mathcal{P}$. Then, to compute the relative difference between two points $\boldsymbol{p}_i$ and $\boldsymbol{p}_j$ and their associated normals $\vec{\boldsymbol{n}}_i$ and $\vec{\boldsymbol{n}}_j$, we define a fixed Darboux coordinate frame at one of the points. In order for the frame to be uniquely defined:

$$\begin{aligned} &\text{if: } \arccos(\vec{\boldsymbol{n}}_i \cdot \vec{\boldsymbol{p}}_{ji}) \leq \arccos(\boldsymbol{n}_j, \vec{\boldsymbol{p}}_{ij}),\ \boldsymbol{p}_{ji} = \boldsymbol{p}_j - \boldsymbol{p}_i,\ \boldsymbol{p}_{ij} = \boldsymbol{p}_i - \boldsymbol{p}_j \\ &\text{then} \begin{cases} \boldsymbol{p}_s = \boldsymbol{p}_i,\ \boldsymbol{n}_s = \boldsymbol{n}_i \\ \boldsymbol{p}_t = \boldsymbol{p}_j,\ \boldsymbol{n}_t = \boldsymbol{n}_j \end{cases} \\ &\text{else} \begin{cases} \boldsymbol{p}_s = \boldsymbol{p}_j,\ \boldsymbol{n}_s = \boldsymbol{n}_j \\ \boldsymbol{p}_t = \boldsymbol{p}_i,\ \boldsymbol{n}_t = \boldsymbol{n}_i \end{cases} \end{aligned} \tag{4.18}$$

where $\boldsymbol{p}_s$ is defined as the source point and $\boldsymbol{p}_t$ as the target. The source point is chosen such that the angle between its associated normal and the line connecting the two points is minimal. The Darboux frame origin (see Figure 4.12) can then be defined at $\boldsymbol{p}_s$ as:

$$\begin{cases} \mathsf{u} = \boldsymbol{n}_s \\ \mathsf{v} = \mathsf{u} \times \dfrac{(\boldsymbol{p}_t - \boldsymbol{p}_s)}{\|\boldsymbol{p}_t - \boldsymbol{p}_s\|_2} \\ \mathsf{w} = \mathsf{u} \times \mathsf{v} \end{cases} \tag{4.19}$$

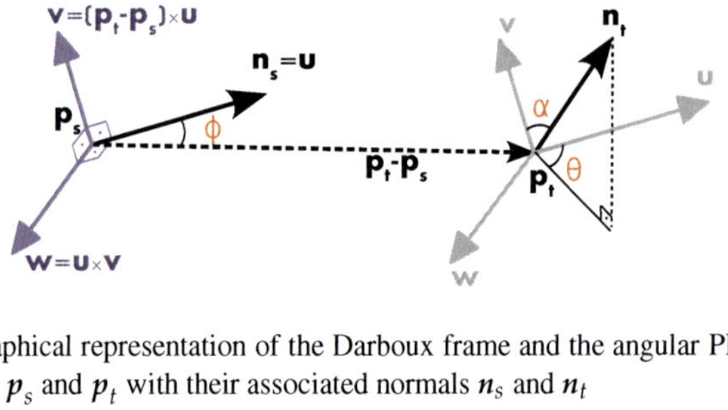

Fig. 4.12 A graphical representation of the Darboux frame and the angular PFH features for a pair of points $\boldsymbol{p}_s$ and $\boldsymbol{p}_t$ with their associated normals $\boldsymbol{n}_s$ and $\boldsymbol{n}_t$

Using the Darboux uvw frame, the difference between the two normals $\boldsymbol{n}_s$ and $\boldsymbol{n}_t$ can be expressed as a set of angular features as follows:

$$\begin{aligned} \alpha &= \mathsf{v} \cdot \boldsymbol{n}_t \\ \phi &= \mathsf{u} \cdot \frac{(\boldsymbol{p}_t - \boldsymbol{p}_s)}{d} \\ \theta &= \arctan(\mathsf{w} \cdot \boldsymbol{n}_t, \mathsf{u} \cdot \boldsymbol{n}_t) \end{aligned} \tag{4.20}$$

where d is the Euclidean distance between the two points $\boldsymbol{p}_s$ and $\boldsymbol{p}_t$, $d = \|\boldsymbol{p}_t - \boldsymbol{p}_s\|_2$. The quadruplet $\langle \alpha, \phi, \theta, d \rangle$ is computed for each pair of two points in the $\mathcal{P}^{k_2}$ neighborhood, therefore reducing the 12 values (x, y, z, n_x, n_y, n_z for each point) of the two points and their normals to 4.

The number of quadruplets formed in a neighborhood $\mathcal{P}^k$ is $k\frac{k-1}{2}$, with an overall theoretical computational complexity of $O(k^2)$. That means that for a point cloud dataset $\mathcal{P}$ with n points, the complexity is $O(nk^2)$.

Figure 4.13 presents an influence region diagram of the PFH computation for a query point ($\boldsymbol{p}_q$), marked with red and placed in the middle of a circle (sphere in 3D) with radius r, and all its k neighbors (points with distances smaller than the radius r) are fully interconnected in a mesh.

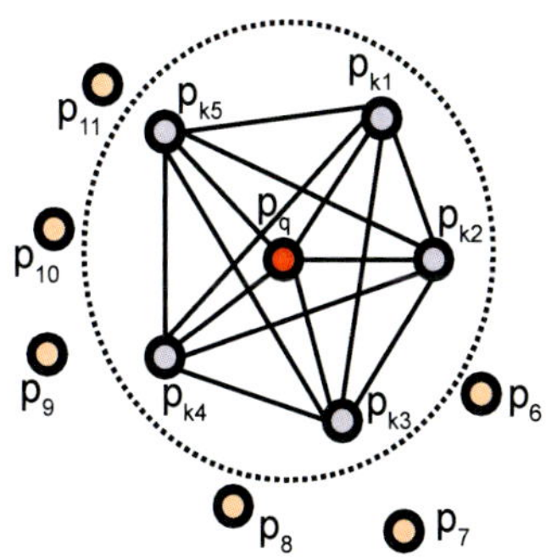

Fig. 4.13 The influence region diagram for a Point Feature Histogram. The query point (red) and its k-neighbors (blue) are fully interconnected in a mesh.

To create the final PFH representation for the query point $\boldsymbol{p}_i$, the set of all quadruplets is binned into a histogram. The binning process divides each features's value range into b subdivisions, and counts the number of occurrences in each subinterval. Since three out of the four features presented above are measure of the angles between normals, their values can easily be normalized to the same interval on the trigonometric circle. A binning example is to divide each feature interval into the same number of equal parts, and therefore create a histogram with b^4 bins in a fully correlated space. In this space, a histogram bin increment corresponds to a point having certain values for all its 4 features [RMBB08b]. Figure 4.14 presents examples of Point Feature Histograms representations for different points in a cloud.

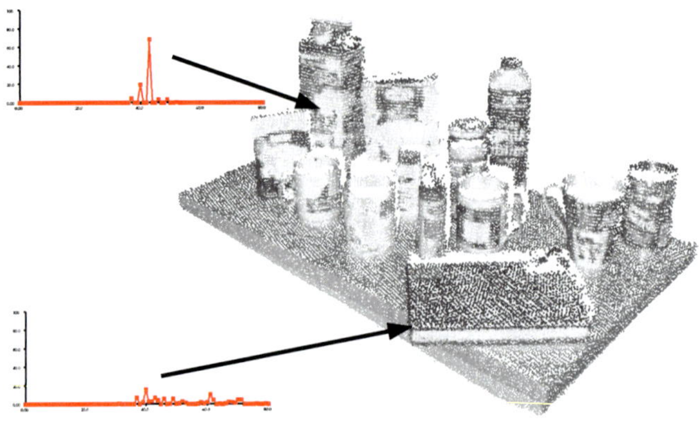

Fig. 4.14 Examples of Point Feature Histograms representations for two selected points in a point cloud dataset

In some cases, the fourth feature, d, does not present an extreme significance for 2.5D datasets, usually acquired in robotics, as the distance between neighboring points increases from the viewpoint. Therefore, omitting d for scans where the local point density influences this feature dimension has proved to be beneficial.

To analyze the discriminating power of the PFH space, an analysis can be performed at how features computed for different geometric surfaces resemble or differ from each other. Figure 4.15 presents the PFH signatures for points lying on 5 different convex surfaces, namely a sphere and a cylinder with a radius of 5 cm, an edge, a corner, and finally a plane. To illustrate that the features are discriminative, in the left part of the figure a confusion matrix is assembled, with gray values representing the distances between the mean histograms of the different shapes, obtained using the Histogram Intersection Kernel [BOV03]:

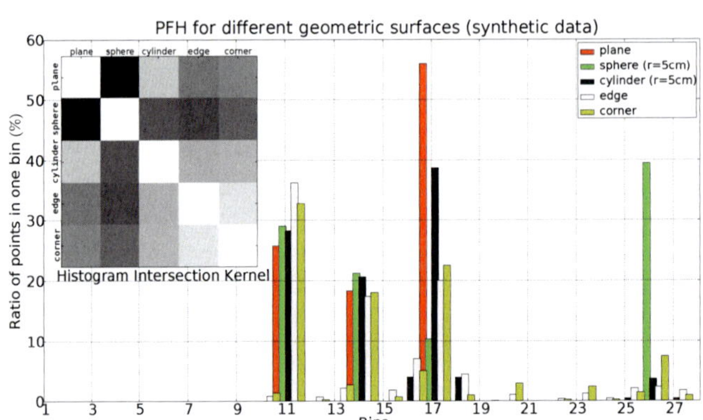

Fig. 4.15 Example of Point Feature Histograms for points lying on primitive 3D geometric surfaces

$$d(PFH_{\mu 1}, PFH_{\mu 2}) = \sum_{i=1}^{b^3} \min(PFH_{\mu 1}^{i}, PFH_{\mu 2}^{i}) \tag{4.21}$$

As shown, points lying on different geometric surfaces produce distinct signatures in the PFH hyperspace.

An interesting aspect regarding the computational complexity of Point Feature Histograms is given by analyzing the number of features that need to be recomputed in the neighborhoods of two query points $\boldsymbol{p}$ and $\boldsymbol{q}$ if $\boldsymbol{p}$ and $\boldsymbol{q}$ are each other's neighbors. In this case, many points from the neighborhood of $\boldsymbol{p}$ will also be part of $\boldsymbol{q}$'s neighborhood, so if their respective histograms are being computed, the temporal locality of data accesses can be exploited using a cache.

Though the theoretical runtime does not improve by using a cache, since all point pairs within a neighborhood still need to be considered, the lookups are considerably faster than the computations of the angular features. However, in order for the cache to be effective in the estimation process, the feature values which have just been computed should be used before they get replaced in the cache. This translates to a requirement that points should be ordered in some manner in the dataset. To illustrate this, Figure 4.16 presents the effects of caching vs standard computations for one unordered (randomized point indices) and one ordered (*i.e.* points that are close in the point cloud should have similar indices) dataset containing the well known Stanford bunny model [TL94].

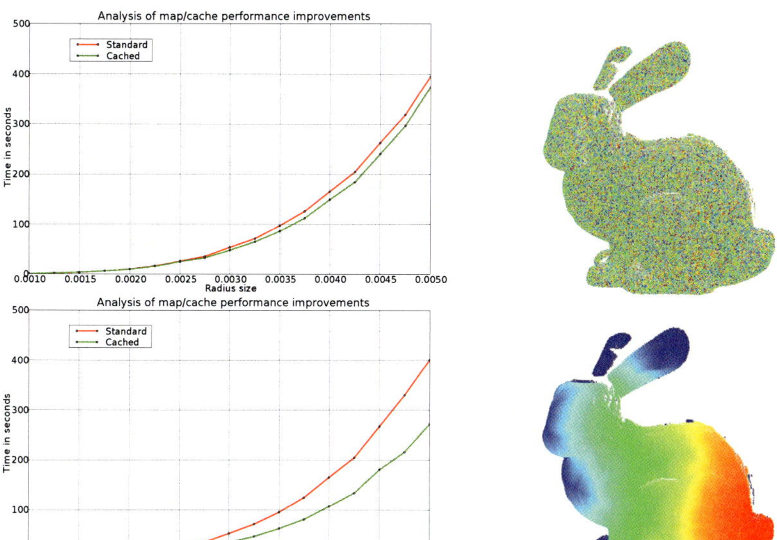

Fig. 4.16 Complexity Analysis on Point Feature Histograms computations for the bunny00 dataset: unordered (top), and reordered (bottom)

The reordering is being performed using a growing algorithm in Euclidean distance space using an octree to achieve similar results as minimal spanning trees in graph theory. The point order is represented in color, ranging from red for low indices to blue for high indices. Note how the cache impacts the PFH computation time considerably for the ordered dataset with a reduced runtime of about 75% compared to the standard computation method. The speedup is lower for the unordered dataset due to the random point order which renders the cache FIFO replacement method suboptimal.

Because PFH signatures depend on the estimated surface normals, high degrees of noise can lead to small variations in the results. If the difference between two similar values is small, but they are on different sides of their respective threshold, the algorithm will place them in different bins. This very interesting effect can easily be observed for planar structures, where estimated surface normals can vary quite a bit in noisy datasets. A solution to this is to make sure that the value intervals for the angular features do not get split in an even number of subdivisions, because that would result in values being placed into two separate bins b_i and b_{i+1} for values $\approx 90°$ (based on whether the value is larger or smaller than $90°$). By selecting an odd number of subdivisions, the resultant histograms become more robust in the presence of additive noise for planes, without influencing the other types of surface primitives.

To illustrate the above, we show the computed feature histograms for a point located in the middle of a planar patch of $10cm \times 10cm$, first on synthetically generated data without noise, and then on data with added zero-mean Gaussian noise with $\sigma = 0.1$ (see Figure 4.18). As shown in Figure 4.17, the estimated histograms are similar even under noisy conditions. Note that the histograms change only slightly as the noise level increases or the radius decreases, thus retaining enough similarity to the generic plane histogram.

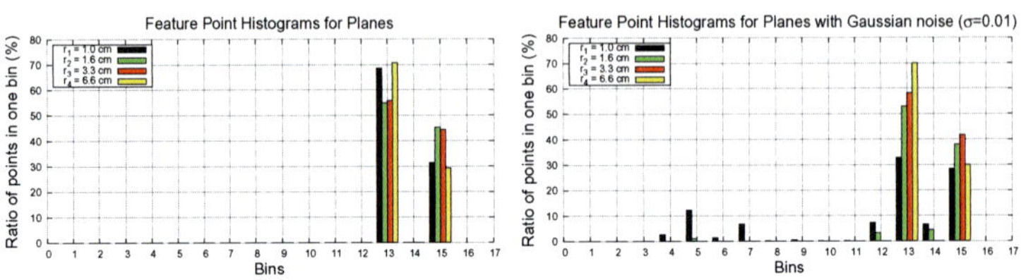

Fig. 4.17 Point Feature Histograms over different radii for a point on a $10cm \times 10cm$ sized plane without noise – left; and with zero-mean Gaussian noise ($\sigma = 0.1$) – right

Fig. 4.18 Point (marked with red) for which the PFH (see Figure 4.17) was computed on a plane without noise – left; and with noise – right. Normals showed for $1/3$ of the points, computed for a radius of 1cm.

4.5 Fast Point Feature Histograms (FPFH)

The theoretical computational complexity of the Point Feature Histogram for a given point cloud $\mathcal{P}$ with n points is $O(nk^2)$, where k is the number of neighbors for each point $\boldsymbol{p}$ in $\mathcal{P}$. As previously shown, a straightforward optimization is to cache feature values and reorder the point cloud dataset so that performing lookups in a data container becomes faster than recomputing the values. For real-time or near real-time applications however, the computation of Point Feature Histograms in dense point neighborhoods can represent one of the major bottlenecks in the mapping framework.

This section proposes a simplification of the PFH formulation, called Fast Point Feature Histograms (FPFH) [RBB09], that reduces the computational complexity of the algorithm to $O(nk)$, while still retaining most of the discriminative power of the PFH.

To simplify the histogram feature computation, we proceed as follows:

- in a first step, for each query point $\boldsymbol{p}_q$ a set of tuples $\langle \alpha, \phi, \theta \rangle$ between itself and its neighbors are computed as described in Equation 4.20 – this will be called the Simplified Point Feature Histogram (SPFH);
- in a second step, for each point its k neighbors are re-determined, and the neighboring SPFH values are used to weight the final histogram of $\boldsymbol{p}_q$ (called FPFH) as follows:

$$FPFH(\boldsymbol{p}_q) = SPFH(\boldsymbol{p}_q) + \frac{1}{k} \sum_{i=1}^{k} \frac{1}{\omega_k} \cdot SPFH(\boldsymbol{p}_k) \tag{4.22}$$

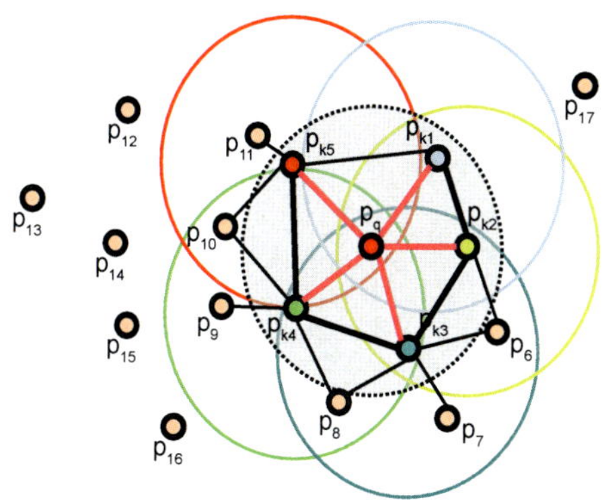

Fig. 4.19 The influence region diagram for a Fast Point Feature Histogram. Each query point (red) is connected only to its direct k-neighbors (enclosed by the gray circle). Each direct neighbor is connected to its own neighbors and the resulted histograms are weighted together with the histogram of the query point to form the FPFH. The thicker connections contribute to the FPFH twice.

where the weight ω_k represents a distance between the query point $\boldsymbol{p}_q$ and a neighbor point $\boldsymbol{p}_k$ in some given metric space, thus scoring the $(\boldsymbol{p}_q, \boldsymbol{p}_k)$ pair, but could just as well be selected as a different measure if necessary. To understand the importance of this weighting scheme, Figure 4.19 presents the influence region diagram for a $\mathcal{P}^k$ set centered at $\boldsymbol{p}_q$.

Thus, for a given query point $\boldsymbol{p}_q$, the algorithm first estimates its SPFH values by creating pairs between itself and its neighbors (illustrated using red lines). This is repeated for all the points in the dataset, followed by a re-weighting of the SPFH values of $\boldsymbol{p}_q$ using the SPFH values of its $\boldsymbol{p}_k$ neighbors, thus creating the FPFH for $\boldsymbol{p}_q$. The extra FPFH connections, resultant due to the additional weighting scheme, are shown with black lines. As the diagram shows, some of the value pairs will be counted twice (marked with thicker lines in the figure). The main differences between the PFH and FPFH formulations are summarized below:

1. The FPFH does not fully interconnect all neighbors of $\boldsymbol{p}_q$ as it can be seen from Figures 4.13 and 4.19, and is thus missing some value pairs which might contribute to capture the geometry around $\boldsymbol{p}_q$;
2. The PFH models a precisely determined surface around $\boldsymbol{p}_q$, while the FPFH includes additional point pairs outside the r radius sphere (though at most $2r$ away);
3. Finally, because of the re-weighting scheme, the FPFH combines SPFH values and re-captures some of the point neighboring value pairs.

A further PFH optimization can be pursued if we tackle the correlation problem in the feature histogram space. So far, the resulting number of histogram bins was given by b^d, where b is the number of *quantum bins* (*i.e.* subdivision intervals in a feature's value range) and d the number of features selected (*e.g.* if $d = 3$ and $b = 5$

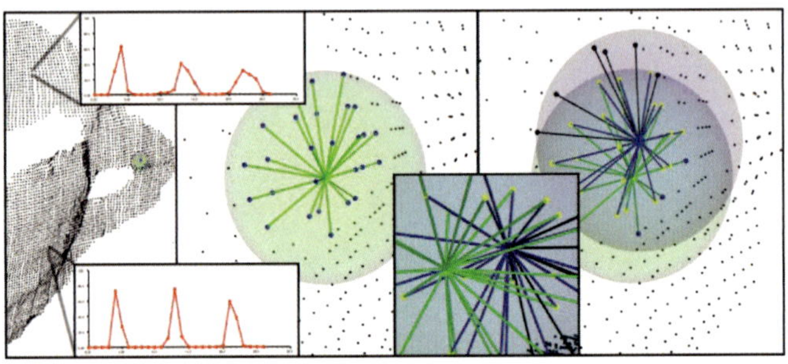

Fig. 4.20 The estimation of a FPFH feature for a 3D point. From left to right: a point $\boldsymbol{p}$ (green) is selected from a dataset, and a set of $\mathcal{P}_k$ neighbors (black) enclosed in a sphere (green) with radius r are selected. For each pair of points $\langle \boldsymbol{p}, \boldsymbol{p}^k \rangle$, $\boldsymbol{p}^k \in \mathcal{P}_k$ (connections showed with green lines), three angular features are computed as presented in Equation 4.20. These steps are repeated for each neighbor as shown in the right part of the figure, and the FPFH of $\boldsymbol{p}$ will be weighted with the estimated features of all its $\boldsymbol{p}^k$ neighbors, as presented in Equation 4.22.

then $5^3 = 125$ bins). This can be described as a subdivided $5 \times 5 \times 5$ cube in 3 dimensions, where one subdivision cell corresponds to a point having certain values for its 3 features, hence leading to a fully correlated feature space. Because of this however, the resulting histogram will contain a lot of zero values (see Figures 4.14 and 4.15), and can thus contribute to a certain degree of information redundancy in the histogram space, as some of the subdivision cells of the cube will never contain any values. A simplification of the above is to decorrelate the values, that is to simply create d separate feature histograms, one for each feature dimension, and concatenate them together. Figure 4.20 presents a graphical description of the FPFH estimation process.

Similar to Figure 4.15, Figure 4.21 presents the FPFH signatures for points lying on the same primitive geometric surfaces. The assembled confusion matrix using the formulation in Equation 4.21 shows that the FPFH signatures are still discriminative with regards to the underlying surface they characterize.

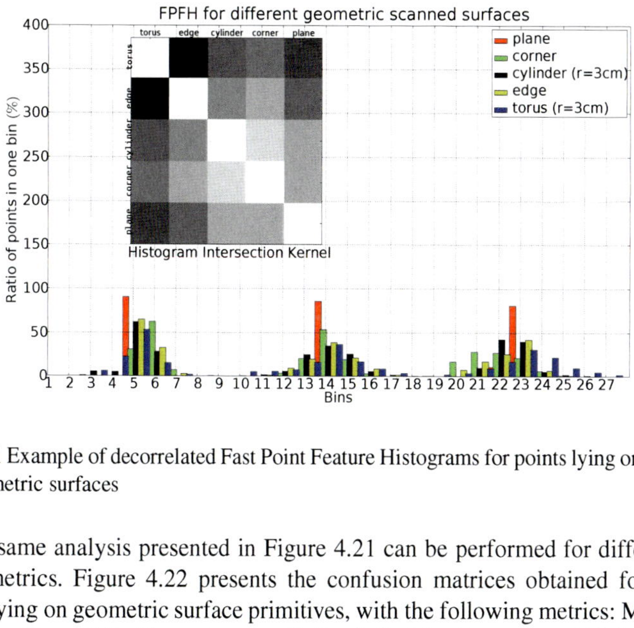

Fig. 4.21 Example of decorrelated Fast Point Feature Histograms for points lying on primitive 3D geometric surfaces

The same analysis presented in Figure 4.21 can be performed for different distance metrics. Figure 4.22 presents the confusion matrices obtained for several points lying on geometric surface primitives, with the following metrics: Manhattan (L1), Euclidean (L2), Bhattacharyya, Kullback-Leibler divergence, Chi-Square, and finally the Histogram Intersection Kernel (see Appendix A.2 for their equations). The discriminative properties of each metric can be extracted from the respective confusion matrix as follows:

$$A_c = 100 \frac{\sum_{i=1}^{N} 1 - |t_i - p_i|}{\sum_{i=1}^{N} p_i} (\%) \tag{4.23}$$

where A_c represents the accuracy of the confusion matrix, p_i represents the actual prediction made, and t_i the true target. Table 4.1 presents the resultant accuracy values for each distance metric used.

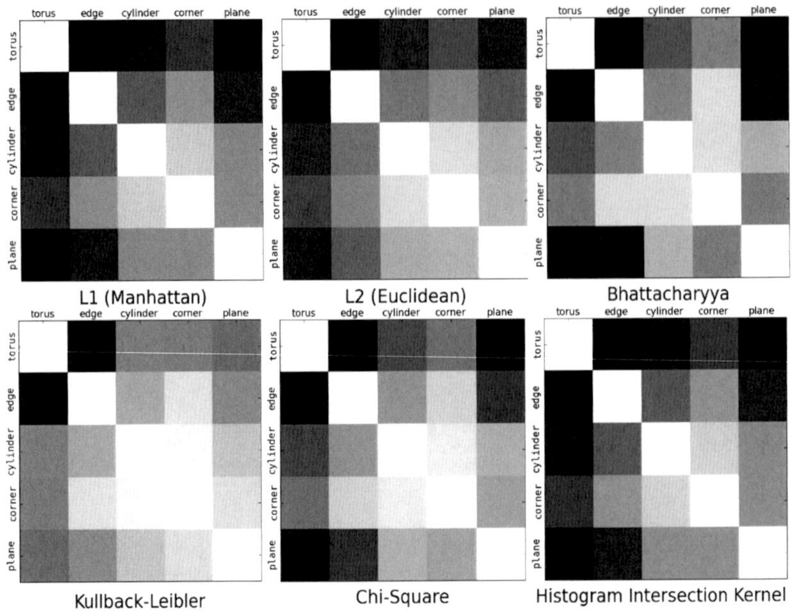

Fig. 4.22 Resultant confusion matrices for the same FPFH representations in Figure 4.21 using various different distance metrics

Table 4.1 Accuracy values for the resultant confusion matrices for FPFH analysis with different distance metrics

Manhattan	Euclidean	Bhattacharyya	Kullback-Leibler	Chi-Square	HIK
42.10 %	37.51 %	37.15 %	30.78 %	35.90 %	42.11 %

4.6 Feature Persistence

When using features as characteristics for a point cloud $\mathcal{P}$, it's important to find a compact subset of points $\mathcal{P}_f$ that best represents the point cloud. The lesser the number of feature points and the better they characterize the data, the more efficient are the subsequent data interpretation steps. However, choosing the subset $\mathcal{P}_f$ is not easy, as it relies on a double dependency: both the number of neighbors k and the point cloud density φ.

The theory of feature persistence or saliency is also connected to the discussion presented in Section 4.1 regarding the *right scale* factor. Given a certain feature representation space and a point $\boldsymbol{p}_q \in \mathcal{P}$, how many neighbors $\boldsymbol{p}_k \in \mathcal{P}^k$ should be selected in order to characterize $\boldsymbol{p}_q$ as best as possible?

One of the solutions adopted by the computer graphics research community [PKG03] is to compute several feature values over multiple scales, and then look for jumps in the feature value curve. In the case of surface normals for example, one can search for jumps in the surface curvature estimation σ_p, as that indicates strong deviations in the normal direction. The critical size of the neighborhood can thus be determined by examining this variation-scale curve.

Unfortunately this assumption only holds true for smooth surfaces, where the value of a certain feature changes monotonically with the neighborhood size. Any point cloud datasets acquired using the techniques mentioned in Section 2.1 are non-smooth and present exacerbated noise levels that lead to large and unforeseen variations in a feature space, when points are being added or removed from $\mathcal{P}^k$.

A solution that we previously proposed in [RMBB08c, RBMB08] for the problem of determining persistent features in a dataset $\mathcal{P}$, is to estimate $\mathcal{P}_f$ based on an analysis of the so called *μ-histogram* of $\mathcal{P}$. In the following, we will show the implementation details of this algorithm and demonstrate that the results obtained are applicable to non-smooth unorganized point cloud datasets.

The purpose of the persistence analysis is to find the most salient points in the data. This is a useful preprocessing step for applications such as point cloud registration. Large datasets having a big number of points with the same or similar feature characteristics could lead to the so called "sliding" problem, where points on certain surfaces contribute negatively to the global distance metric due to ambiguous correspondences [GIRL03]. A solution would be to either segment out these surfaces at an a priori step, or to neglect all points with features which are considerably dominant in the dataset and thus concentrate on more prominent (salient) points.

To select the best set of neighbors $\mathcal{P}^k$ for a query point $\boldsymbol{p}_q$, the proposed method varies the size of the neighborhood by enclosing $\boldsymbol{p}_q$ in a sphere of radius r_i having $\boldsymbol{p}_q$ as its center. The sphere radius takes multiple values between some $r_{\min}$ and $r_{\max}$ values, depending on the point cloud size and density. Given that the feature representation of choice is a Fast Point Feature Histogram, for each neighborhood $\mathcal{P}_i^k$, a local FPFH is estimated for $\boldsymbol{p}_q$. Then, all the points in $\mathcal{P}$ are selected and their feature representations are used to estimate the *μ-histogram* of $\mathcal{P}$ as the mean of the feature distribution across all Fast Point Feature Histograms.

The FPFH selection criterion at a given scale is motivated by the fact that in a given metric space, one can compute the distances from the mean FPFH (*μ-histogram*) of a dataset to all the features of that dataset. The resultant distances can be used to build a distribution that can be approximated with a Gaussian having a mean μ_d and a standard deviation σ_d. As shown in [RBMB08], by performing a statistical analysis over this distance distribution, features whose values are outside the $\mu_d \pm \beta \cdot \sigma_d$ interval can be selected as less common (therefore *unique*). The parameter β controls the width of the interval and acts as a band-stop filter cut-off parameter.

To account for density variations but also different scales, the above is repeated over each r_i, and *unique* points over the entire interval are marked as *persistent*. In particular, a point $\boldsymbol{p}_i$ is persistent if: i) its FPFH is selected as *unique* with respect to a given radius; and ii) its FPFH is selected in both r_i and r_{i+1}, that is:

$$\mathcal{P}_f = \bigcup_{i=1}^{n-1} [\mathcal{P}_{f_i} \cap \mathcal{P}_{f_{i+1}}] \tag{4.24}$$

where $\mathcal{P}_{f_i}$ represents the set of points which are selected as unique for a given radius r_i, and n is the number of radii subdivisions chosen.

The values of the r_i radii set are selected based on the size of the features that need to be detected. Based on the sensor's resolution, the neighborhood can contain anything from a few points to thousands or even more. For most datasets, fixing the value of β between $[1;2]$ will give satisfactory results.

Because the FPFH space is hyper-dimensional, selecting the most appropriate distance metric can have a big influence on the creation of the distance distribution. Though a certain distance metric is only relevant for one particular point feature representation, it is important to discover this metric through an offline analysis. For the purpose of evaluating the persistence of FPFH representations, we evaluated the following metrics: Manhattan (L1), Euclidean (L2), Jeffries-Matusita (JM), Bhattacharyya (B), Chi-Square (CS), and Kullback-Leibler (KL) [RMBB08c]

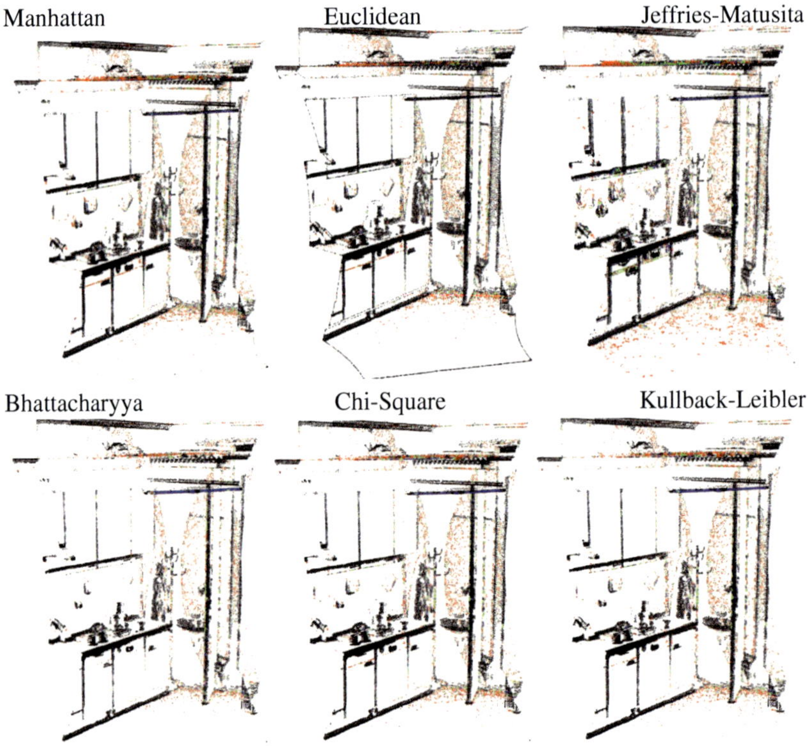

Fig. 4.23 Analyzing the persistence of PFH features in a point cloud for four different radii (r_1 - red, r_2 - green, r_3 - blue, r_4 - black, with $r_1 < r_2 < r_3 < r_4$) using different distance metrics

(see Appendix A.2 for their equations). An example for a dataset acquired in a kitchen environment is given in Figures 4.24. The resultant unique points are selected using the statistical principle described above, where their distance to the *μ-histogram* of the dataset is computed in a given metric space. Figure 4.23 shows them with different colors for four different selected radii. Notice that some metrics such as the Euclidean distance also select the points near the edges of the dataset as unique, which is undesirable for correspondence analysis.

Using the persistence analysis algorithm, the most unique points over the FPFH space can be selected for an interval of radii, here using the Kullback-Leibler divergence, as shown in the middle part of Figure 4.24. The right part presents the overall persistent points in the dataset in blue color, selected using Equation 4.24. Similar examples for different datasets are given in our previous work [RBMB08, RBB09].

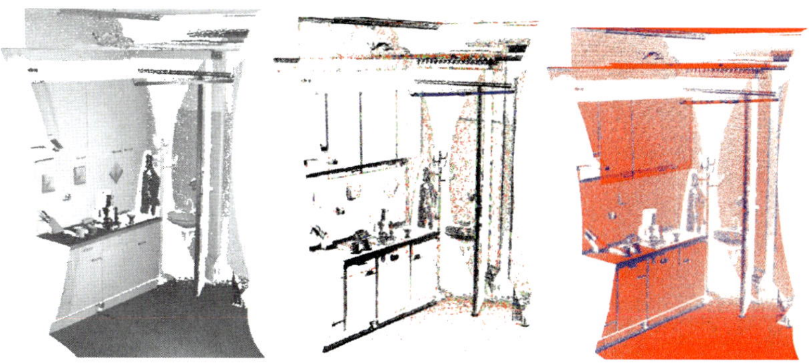

Fig. 4.24 From left to right: raw data shown in monochrome intensity scale (left); uniquely selected points for four different radii (middle); overall persistent points shown with blue color (right)

The left part of Figure 4.25 illustrates the differences between query points located on various geometric surfaces, in a PFH 16D feature hyperspace. The surfaces were synthetically generated to have similar scales, densities, and noise levels as our real-world datasets. For each of the mentioned surfaces, a point was selected such that it lies: (i) on a plane, (ii) on the middle of an edge of a cube (normals pointing both inside and outside), (iii) on the lateral surface of a cylinder at half the height, and (iv) on a sphere. The PFH representations were generated using all the $\mathcal{P}^k$ neighbors inside a sphere with radius r of 2cm. The right part of the figure shows the *μ-histogram* of the kitchen dataset presented in Figure 4.24. As shown, the *μ-histogram* of the dataset resembles a lot the one of a planar surface. This proves an important assumption about the planarity of datasets acquired in indoor environments.

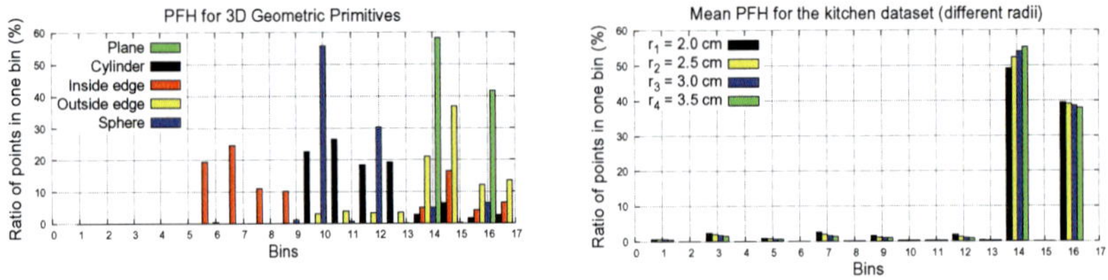

Fig. 4.25 PFH for query points located on different geometric surfaces (left). Mean PFH over different radii for the kitchen scene from Figure 4.24 (right).

4.7 Related Work

In general, the consensus is that while some feature representations are better than others for a particular problem, there is no clear "winner takes it all" representation that surpasses all the other similar approaches. In this section we tackle some of the related research initiatives used in similar applications such as ours.

The local (3D point level) feature estimation problem has been under investigation for a long time in various research fields, such as computer graphics, robotics, and pattern matching, see [TV04, JD00, BP06] for comprehensive reviews. Two basic geometric point features are represented by the underlying surface's estimated curvature and normal [AA04], which have been also investigated for noisier point datasets [MN03]. Other proposals for simple point features include moment invariants [SH80], and integral volume descriptors [GMGP05]. They have been successfully used in applications such as point cloud registration, and are known to be translation and rotation invariant, but they are not discriminative enough for determining the underlying surface type. Specialized point features that detect peaks, pits and saddle regions from Gaussian curvature thresholding for the problem of face detection and pose estimation are presented in [AU07].

In general, descriptors that characterize points with a single value are not expressive enough to make the necessary distinctions for point-surface classification or for correspondence search. As a direct consequence, most scenes will contain many points with the same or very similar feature values, thus reducing their informative characteristics.

Some of the more complex 3D point feature extraction approaches include: spherical harmonic invariants [BH95], spin images [JH99], curvature maps [GGGZ05], or more recently, conformal factors [MC08]. Spherical harmonic invariants and spin images have been successfully used for the problem of object recognition for densely sampled datasets, though their performance seem to degrade for noisier and sparser datasets [BP06]. Conformal factors are based on conformal geometry, which is invariant to isometric transformations, and thus obtains good results on databases of watertight models. Its main drawback is that it can only be applied to manifold meshes. Curvature maps have only been studied in the context of local shape comparisons for data registration. A different point fingerprint representation using the projections of geodesic circles onto the tangent plane at a point p_i was proposed in [SA01] for the problem of surface registration. As the authors note, geodesic distances are more sensitive to surface sampling noise, and thus are unsuitable for real sensed data without a priori smoothing and reconstruction. A decomposition of objects into parts learned using spin images is presented in [HKDH04] for the problem of vehicle identification. However, most of the initiatives presented here require densely sampled data and are not always able to deal with the amount of noise usually present in 2.5D scans.

Following the work done by the computer vision community which developed reliable 2D feature descriptors, some preliminary work has been done recently in extending these descriptors to the 3D domain. In [AAvd07], a keypoint detector called THRIFT, based on 3D extensions of SURF [BETG08] and SIFT [Low04b], is used to detect repeated 3D structures in range data of building facades, however the authors do not address computational or scalability issues, and fail to compare their results with the state-of-the art. A 3D descriptor based on SIFT is used in [PSM07] to capture spatio-temporal shapes, for the purpose of action recognition from video sequences. The classification of a sequence is obtained by randomly picking 200 points and the results are encouraging, but it is unclear whether they can be extended to work reliably for large and noisy point cloud datasets. In [SS07], the authors propose RIFT, a rotation invariant 3D feature descriptor utilizing techniques from SIFT, and perform a comparison of their work with spin images, showing promising results for the problem of identifying corresponding points in 3D datasets.

In general it is not clear how one should select the optimal k support for a point when computing most of the above mentioned features. If the data is highly contaminated with noise, selecting a small k will lead to large errors in the feature estimation. However, if k is too big, small details will be suppressed. Recently, work has been done on automatic computation of good k values (*i.e. scale*) for normal estimation on 3D point cloud data [MN03, LUVH05] as well as principal curvatures [PKG03, YLHP06, KSNS07] on multiple scales. Unfortunately, some of the above mentioned methods for computing an optimal scale require additional thresholds, such as d_1 and d_2 which are determined empirically in [MN03], and estimated using linear least-squares in [LUVH05] when knowledge about ground truth normals exists. In [PKG03] the neighborhood is grown incrementally until a jump occurs in the variation-scale curve, but the method cannot be successfully applied to noisy point clouds, as the variations in the surface curvature are not modified smoothly with k. The selection of the T_c threshold in [YLHP06] is not intuitive, and the authors do not explain properly if the resulted persistent features are obtained using solely the intersection of features computed over different radii. The statistical estimation of curvatures in [KSNS07] uses a M-estimation framework to reject noise and outliers in the data and samples normal variations in an adaptively reweighted neighborhood, but it is unfortunately slow for large datasets, requiring approximately 20 minutes for about 10^6 points.

Methods relying on global features include descriptors such as Extended Gaussian Images (EGI) [Hor84], eigen shapes [CF], or shape distributions [OFCD02]. The latter samples statistics of the entire object and represents them as distributions of shape properties, however they do not take into account how the features are distributed over the surface of the object. Eigen shapes show promising results but they are viewpoint dependent and cannot generalize to complete 3D object classes. EGIs describe objects based on the unit normal sphere, but have problems handling arbitrarily curved objects.

4.8 Summary

This chapter presented novel contributions to the problem of informative 3D point feature representations from noisy datasets. Both PFH and FPFH are shown to be discriminative 3D feature descriptors, useful for the problem of correspondence search and point classification with respect to the underlying geometric surface primitive the point is lying on. The persistence of selected unique histograms has been analyzed using different distance metrics over multiple scales, and a subset characterizing the input dataset has been selected as representative for a given point cloud dataset.

Chapter 5
From Partial to Complete Models

> *'The person who can combine frames of reference and draw connections between ostensibly unrelated points of view is likely to be the one who makes the creative breakthrough."*
>
> DENISE SHEKERJIAN

This chapter presents mechanisms for the integration of partial datasets acquired from different views into a consistent global model. In particular, we will address the problems of: i) point cloud *registration*, to transform and align scans into the same coordinate system, and ii) data *resampling*, to create smooth datasets and filter low frequency point outliers.

5.1 Point Cloud Registration

The problem of consistently aligning various 3D point cloud data views into a complete model is known as *registration*. Its goal is to find the relative positions and orientations of the separately acquired views in a global coordinate framework, such that the intersecting areas between them overlap perfectly. For every set of point cloud datasets acquired from different views, we therefore need a system that is able to align them together into a single point cloud model, so that subsequent processing steps such as segmentation and object reconstruction can be applied. The work presented herein is motivated by finding correct point-to-point correspondences in real-world noisy data scans, and estimating rigid transformations that can rotate and translate each individual scan into a consistent global coordinate framework.

A motivation example in this sense is given in Figure 5.1, where a set of six individual datasets has been acquired using a tilting 2D laser unit. Since each individual scan represents only a small part of the surrounding world, it is imperative to find ways to register them together, thus creating the *complete* point cloud model shown in the left part of Figure 5.2. Once this model is obtained, we can make use of segmentation and model fitting techniques to construct representations such as the ones shown in right part of the figure.

R.B. Rusu: *Semantic 3D Object Maps for Everyday Robot Manipulation*, STAR 85, pp. 61–74.
DOI: 10.1007/978-3-642-35479-3_5

Fig. 5.1 Six individual point cloud datasets acquired from different views

Fig. 5.2 Left: The registration of the six datasets from Figure 5.1 into a consistent and complete point cloud model. Right: A segmentation example describing furniture candidates identified in the data.

The registration problem becomes easily solvable if point to point correspondences are perfectly known in the input datasets. This means that a selected list of points $\boldsymbol{p}_i \in \mathcal{P}_1$ have to "coincide" from a feature representation point of view with another list of points $\boldsymbol{q}_j \in \mathcal{P}_2$, where $\mathcal{P}_1$ and $\mathcal{P}_2$ represent two partial view datasets. Differently said, the sets of $\boldsymbol{p}_i$ and $\boldsymbol{q}_j$ have been sampled on the same real world surfaces, but from different acquisition poses. This means however, that in the complete point cloud model that needs to be created, they could be merged together, especially if their coordinates are equal $\boldsymbol{p}_i = \boldsymbol{q}_j$, and thus reduce the number of points overall. Because the quality of the datasets is influenced by sensor noise and other perturbing factors, the coordinates of the points will almost never be equal unfortunately, and the above simplification will not hold.

One of the most popular registration methods for unorganized point cloud datasets to date is the Iterative Closest Point (ICP) algorithm [BM92, Zha], an iterative descend method which tries to find the optimal transformation between two datasets by minimizing the Euclidean distance error metric between their overlapping areas. ICP uses pairs of nearest 3D points in the source and model set as correspondences, and assumes that every point has a corresponding match. Since for the application presented herein, it would be impossible to presume that every point $\boldsymbol{p}_i$ scanned in $\mathcal{P}_1$ has a known correspondence in $\boldsymbol{q}_j \in \mathcal{P}_2$, it becomes essential to create mechanisms that can robustly estimate good correspondences.

Unfortunately, ICP has other additional drawbacks, such as being susceptible to local minima, having a small convergence basin, and in general needing a high number of iteration steps until convergence can be reached. Though there have been a lot of efforts into improving the correspondence and registration problem (see Section 5.3 for a discussion on similar related initiatives), there is no single approach that can solve all these problems for any particular set of input point clouds used.

The contributions presented in this chapter tackle the point correspondence problem, for which we propose the use of our feature representations as presented in Sections 4.4 and 4.5. Additionally, we will construct geometric constraints between correspondent representations in different datasets, that will help avoid the algorithm being trapped into local minima solutions.

Thus, the first step of our proposed registration module is to compute a good initial alignment, and return a set of m correspondences that can be used to directly transform the source point cloud into the convergence area of the target. Figure 5.3 presents an example for two partial views of the kitchen dataset from the list presented in Figure 5.1, where a set of good point correspondences has been estimated and a simple rigid transformation has been computed so that the two datasets can be aligned in the same coordinate system. Notice that based on the quality of the correspondences found, the overlapping areas of the two input datasets are almost coincidental after this transformation, which can serve as a good initial guess for fine tuning the registration using algorithms such as ICP.

To improve the chances of point feature representations being similar in datasets acquired from different poses, the proposed registration algorithm first performs an analysis of the persistence of a feature using techniques similar to the ones presented in Section 4.6. Then, we proceed at estimating sets of geometric constraints between corresponding feature representations (here, PFH) in two different datasets. The estimated correspondences must rely on the intrinsic, rotation and position invariant properties of the point cloud. Point to point distances within the point clouds are a good candidate since they are easy to compute and fulfill this requirement, so the algorithm identifies a set of m points $\boldsymbol{p}_i$ that have the same distances to each other than their corresponding points $\boldsymbol{q}_j$.

In a more general sense, the goal is to create a formulation which:

- can transform the source dataset from an arbitrary initial position with respect to the model, to a "close enough" solution (we will call this an *Initial Alignment* method);
- can refine the solution found by the Initial Alignment algorithm from "close enough" to a global optimum.

The above formulation represents a coarse to fine registration mechanism. Since the latter step is well known (see Section 5.3) and can be parameterized correctly in many ways, the contributions presented in this chapter employs methods that help the first step, namely a greedy Initial Alignment similar to the one presented in [GMGP05], and a sample consensus based Initial Alignment algorithm, as published in our previous work [RBB09].

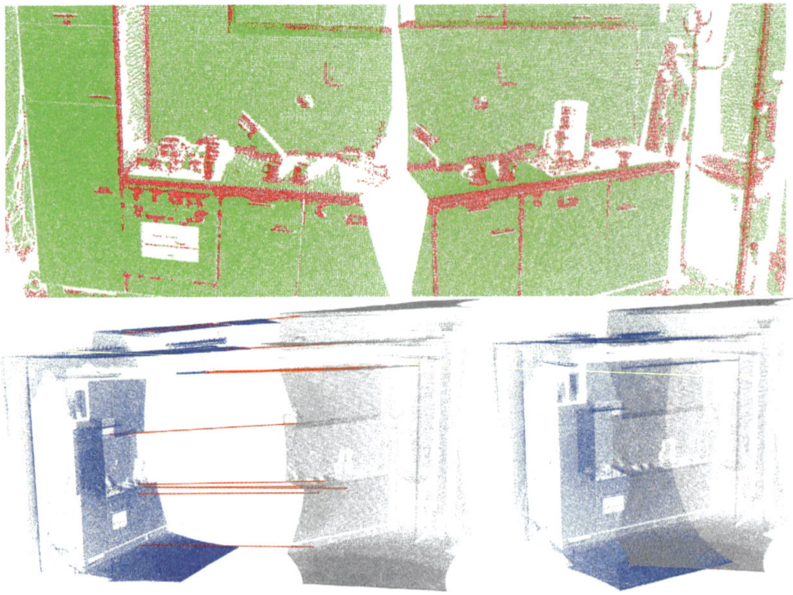

Fig. 5.3 Aligning two scans of a kitchen dataset using a set of good PFH correspondences. The persistent PFH representations selected as interesting points for registration are shown at the top.

In the first step of the greedy Initial Alignment method, the algorithm searches for a set of corresponding points in the target point cloud for the persistent features of the source. To measure the similarity of two point feature representations, we compute the distance between the PFH using an appropriate distance metric as described in detail in [RBMB08, RMBB08c]. To do this efficiently, a n-dimensional kD-tree is created in the PFH feature histograms space, where n represents the number of PFH bins used, and for each persistent feature point in the source dataset, a k-nearest neighbor search is performed in the target. Hence, we look at the k points with the most similar histograms and keep them as correspondence candidates for $\boldsymbol{p}_i$. We call this set of correspondence candidates:

$$C = \{c_i \| c_i = \langle \boldsymbol{p}_i, \boldsymbol{q}_{i1}, ..., \boldsymbol{q}_{ik} \rangle, 0 < i < m\} \tag{5.1}$$

In the second phase of the greedy Initial Alignment, the best entries $\langle c_i, c_j \rangle$ from C are merged into a set of 2-point correspondences E_2. For every combination of $\boldsymbol{p}_i$ and $\boldsymbol{p}_j$, the goal is to find two points $\boldsymbol{q}_{ir}$ and $\boldsymbol{q}_{js}$ that minimize the quantity:

$$||\boldsymbol{p}_i - \boldsymbol{p}_j| - |\boldsymbol{q}_{ir} - \boldsymbol{q}_{js}|| \tag{5.2}$$

and add them to E_2, in increasing order. The method then proceeds at merging pairs of entries $\langle e_{2,i}, e_{2,j}\rangle$ from E_2 if $e_{2,i}$ is the next entry that is not merged yet. If:

$$\mathcal{P}^* = \{\boldsymbol{p}_i^* \in e_{2,i} \cup e_{2,j}\} \tag{5.3}$$

and

$$\mathcal{Q}^* = \{\boldsymbol{q}_i^* \in e_{2,i} \cup e_{2,j}\} \tag{5.4}$$

then $e_{2,j}$ minimizes $dRMS(\mathcal{P}^*, \mathcal{Q}^*)$, where:

$$dRMS^2(\mathcal{P}^*, \mathcal{Q}^*) = \frac{1}{n^2}\sum_{i=1}^{n}\sum_{j=1}^{n}(\|\boldsymbol{p}_i^* - \boldsymbol{p}_j^*\| - \|\boldsymbol{q}_i^* - \boldsymbol{q}_j^*\|)^2, \tag{5.5}$$

analog to the construction of E_2. The system continues merging entries from E_{2^k} into $E_{2^{k+1}}$ in this manner, until there are not enough correspondences left in $E_k(k = k_{max})$ to generate $2k$-point correspondences. The result is a set of $2^{k_{max}}$ points in the source point cloud, each with a corresponding point in the target model such that the point-to-point distances within each of the two sets differ minimally. This $2^{k_{max}}$-point correspondence can be used to construct a transformation (translation and rotation) that aligns these points and is then applied to the source point cloud to roughly align it onto the target.

In the second stage of registration, our variant of ICP using instantaneous kinematics, converges faster to the correct solution than regular ICP. This is mostly due to the point-to-surface error metric which doesn't restrict tangential movement, thus accelerating convergence when the point clouds are already close to each other. This has no closed-form solution, but its approximation has better convergence properties than regular ICP [PLH04] (quadratic vs linear convergence).

Thus after an initial alignment has been obtained from this last set of correspondences, in order to improve convergence speed, the algorithm uses a different method to estimate the optimal transform than classical ICP. We are using an alternative error function that instead of minimizing the sum of squares of distances between corresponding points:

$$\min \sum_{i=1}^{n} dist(R \cdot \boldsymbol{p}_i + T, q_i) \tag{5.6}$$

uses an approximate measure of point-to-surface distances [PLH04].

Using the new distance measure, the classical ICP error function:

$$\sum_{i=1}^{n} \|R \cdot \boldsymbol{p}_i + T - \boldsymbol{q}_i\|^2 \tag{5.7}$$

changes to:

$$\sum_{i=1}^{n} \|(R \cdot \boldsymbol{p}_i + T - \boldsymbol{q}_i) \cdot \vec{\boldsymbol{n}}_{\boldsymbol{q}_i}\| \tag{5.8}$$

where $\vec{n}_{q_i}$ is the normal to the surface in point $\boldsymbol{q}_i$. This means we try to minimize the distance between a point $\boldsymbol{p}_i$ and the surface of its corresponding point $\boldsymbol{q}_i$, or the distance along the normal of $\boldsymbol{q}_i$.

Several experiments have been conducted to evaluate the proposed registration method. Computing the initial alignment using PFH representations proved to be far superior to using other features such as Integral Volume Descriptors (IVD) [GMGP05] or moment invariants [SH80], since it was able to robustly bring the point clouds close to the correct position independent of their original poses (see Figure 5.4). We could only achieve this using IVD when we considered at least $k = 150$ similar points in the target point cloud for every selected feature point in the source, and even then there were cases where it failed. Also, runtime increases exponentially in k, which also renders the IVD method inferior. We empirically discovered that $k = 10$ is a good value for the number of correspondent candidates, as increasing it further did not improve the results.

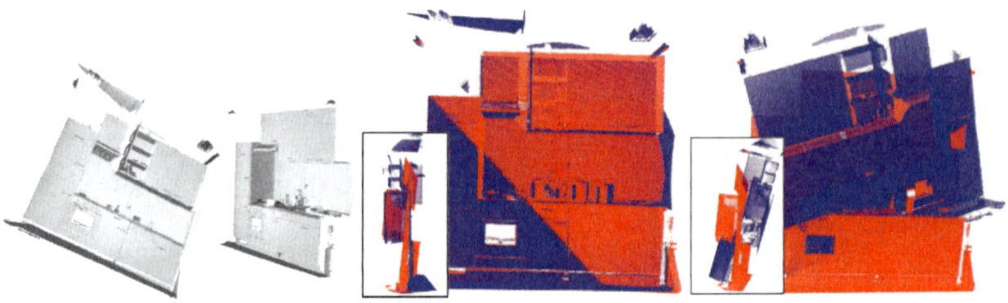

Fig. 5.4 Left: two datasets before registration (remission values shown in grayscale). Middle and right: Initial Alignment results using PFH representations and using IVD.

The greedy Initial Alignment method described above is very robust since it uses point cloud intrinsic, rotation invariant features. However, the computational complexity is relatively high since the merging step requires to look at all possible correspondence pairs. Also, since it is a greedy method, there are situations where it might get trapped in a local minimum. Therefore, we propose another Initial Alignment method in the following, based on a sample consensus formulation that tries to maintain the same geometric relations of the correspondences without having to try all combinations of a limited set [RBB09]. Instead, we sample large numbers of correspondence candidates and rank each of them very quickly by employing the following scheme:

1. Select s sample points from $\mathcal{P}$ while making sure that their pairwise distances are greater than a user-defined minimum distance $d_{\min}$;
2. For each of the sample points, find a list of points in $\mathcal{Q}$ whose histograms are similar to the sample points histograms. From these, select one randomly which will be considered that sample points' correspondence;

3. Compute the rigid transformation defined by the sample points and their correspondences and compute an error metric for the point cloud that computes the quality of the transformation.

The random selection of the correspondences in the second step might seem suboptimal. However, it is a decision that can be made very quickly, where as a more complicated strategy that tries to maximize histogram similarity *and* distance similarities might very likely require multiple passes over the lists of correspondence candidates.

We conducted several experiments to determine a good error metric for the third step. The simplest approach is to count the number of points from $\mathcal{P}$ that fall within an ϵ-band of $\mathcal{Q}$. A slightly more complicated penalty measure applies the Huber loss function L_h:

$$L_h(e_i) = \begin{cases} \frac{1}{2}e_i^2 & ||e_i|| \leq t_e \\ \frac{1}{2}t_e(2||e_i|| - t_e) & ||e_i|| > t_e \end{cases} \tag{5.9}$$

These three steps are repeated, and the transformation that yielded the best error metric is stored and used to roughly align the partial views. Finally, a nonlinear local optimization is applied using a Levenberg-Marquardt algorithm similar to [Fit01]. Figure 5.5 presents results obtained using the sample consensus Initial Alignment algorithm for two datasets acquired in an indoor office environment. The combinatorial nature of the greedy Initial Alignment method makes it extremely slow for large datasets such as these, and the only possible solution to speed it up is to downsample the input data. However this results in "averaged" feature representations which lose the fine details of the sampled surface geometry. The sample consensus Initial Alignment does not suffer from these shortcomings, and outperforms the greedy version in almost every way [RBB09].

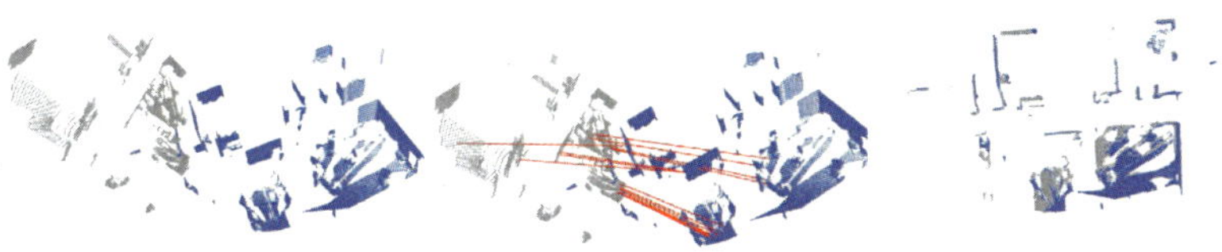

Fig. 5.5 Left: two datasets before registration, middle: sample consensus Initial Alignment correspondences, right: results after registration

The left part of Figure 5.2 presents the six successfully aligned point cloud datasets using the proposed registration approach. The validation of the methods presented herein on different outdoor or synthetic datasets have already been presented in [RMB+08b, RBB09]. The proposed registration methods can also be applied to datasets acquired at different time intervals from similar positions. Figure 5.6 shows a snapshot of the alignment process for two point clouds taken at different instances in time, together with the registration results. Here the robot observed the changes in the region of a cuboid, detecting the difference produced by a cupboard door being opened. This information can also be used to verify some assumptions about the cupboard's location and opening direction.

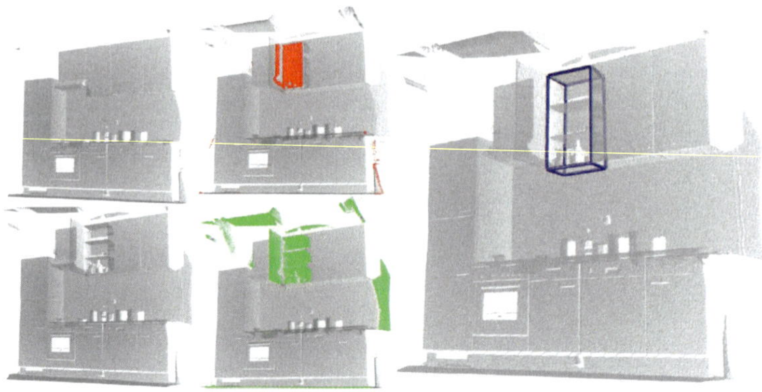

Fig. 5.6 From top to bottom and left to right: two scans of the kitchen scene depicting two states of a cupboard (closed and open), the highlighted temporal differences in both scans and the segmented container

5.2 Data Resampling

Another form of noise filtering and surface smoothing is represented by the resampling category of techniques. In contrast to the previously presented filtering approaches from Section 4.2, where the overall density of the resultant filtered point cloud data set $\mathcal{P}^*$ is approximatively similar, if not equal, to the one of the input cloud $\mathcal{P}$, resampling methods usually attempt to suppress outliers through averaging or interpolation.

To motivate the need for resampling, the left part of Figure 5.7 presents the resultant complete point cloud model $\mathcal{P}_{12}$ after the registration of the two kitchen datasets $\mathcal{P}_1$ and $\mathcal{P}_2$ from Figure 5.3. Because the overlapping area $\mathcal{O}$ of $\mathcal{P}_1$ and $\mathcal{P}_2$ is almost coincident, the density of the points in that area is approximatively two times higher than the rest of the dataset. This variable point density will have a negative effect on the estimation of surface curvatures or normals at each point $\boldsymbol{p}_i \in \mathcal{O}$. The left part of the figure thus shows an attempt to estimate surface curvatures and normals before resampling, while the right part presents the resultant values after resampling for a small planar patch selected from the overlapping area $\mathcal{O}$.

The classical resampling methods made their way into robotics applications from the computer graphics research community [Lev03], and are usually formulated as

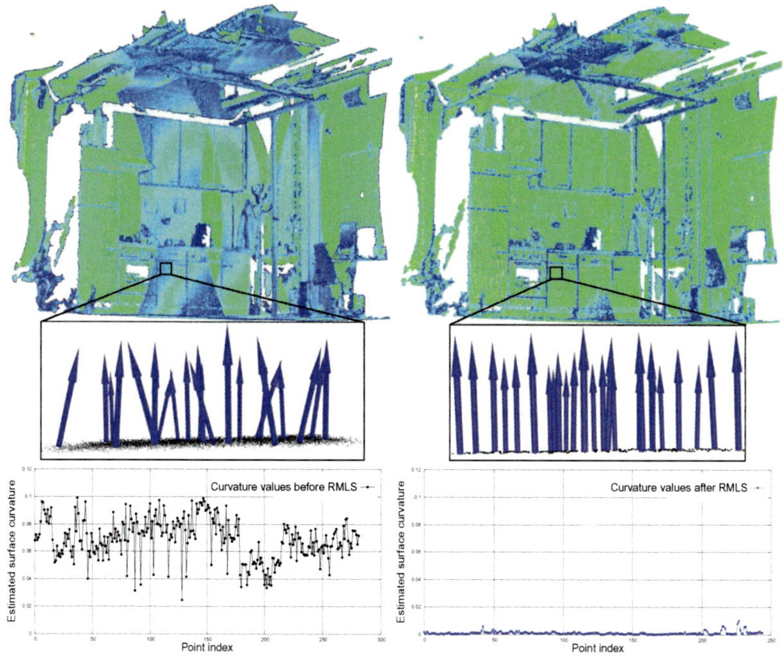

Fig. 5.7 Surface curvatures and normals before and after resampling for two registered kitchen datasets. The estimated values are shown in logarithmic scale for: the global point cloud model after registration (left); the resampled point cloud model (right). Notice the large errors in the surface normal estimates. The plots below depict the estimated surface curvatures before and after resampling.

Moving Least Squares (MLS) solutions. A MLS surface provides an interpolating surface for a given set of points $\mathcal{P}$ by fitting higher order bivariate polynomials to each point neighborhood locally. In contrast to other interpolation or resampling techniques, MLS has the advantage that the resultant fitted surface passes through the original data points.

Besides variable point densities, a complete point cloud model could have regions which are "doubled" in overlapping areas, due to small registration errors, and holes resulted due to scanning surfaces which did not return any distance measurements (*e.g.* shiny, metallic objects). These issues are of extreme importance with respect to the quality of the subsequent geometric processing steps, such as segmentation or surface classification, and need to be alleviated where possible. To illustrate some of these artifacts, the left part of Figure 5.8 presents a smaller dataset representing a mug, while the middle and right parts of the figure present the downsampling and upsampling results using the algorithm presented in this section.

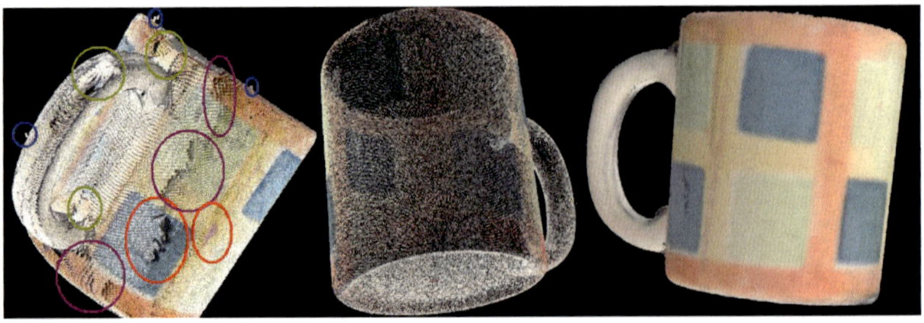

Fig. 5.8 Examples of point cloud data downsampling (middle) and upsampling (right) for a given complete point cloud dataset with errors (left), such as: measurement errors (red), registration errors (purple), holes (brown), and outliers (blue)

To smooth out these anomalies, we proceed as follows. Given a point cloud dataset $\mathcal{P}$, we want to reproduce the complete smooth surface it represents, by resampling (either upsampling or downsampling) and discarding unwanted data. The first step normalizes the coordinates of $\mathcal{P}$ using the diagonal of the point cloud's bounding box, ensuring a unity maximal distance between points. The weight factor is estimated as:

$$h = \mu_d + k \cdot \sigma_d \tag{5.10}$$

where μ_d is the mean and σ_d the standard deviation of the distribution of mean distances between points, and k is a user chosen factor. Then, an initial guess is computed by approximating the point cloud with a set of points $\mathcal{Q}$, sampled in the vicinity of the original points $\mathcal{P}$. To resample the surface as evenly as possible, $\mathcal{Q}$ is represented as a set of equidistant grid points, generated in the proximity of the original dataset.

To fill small holes in $\mathcal{P}$, extra points are added to areas where a hole is detected. The hole detection algorithm performs an analysis of the angles a certain point forms with respect to its neighbors. If $\Theta = \{\theta_1 \cdots \theta_n\}$ is the list of sorted angles between the lines formed by two points $\boldsymbol{p}_{k_1}$ and $\boldsymbol{p}_{k_2}$ in the neighborhood $\mathcal{P}^k$ with the query point $\boldsymbol{p}_q$, then $\boldsymbol{p}_q$ is a point located on the boundary of a hole, given:

$$\max(\alpha = \theta_{i+1} - \theta_i) \geq \alpha_{th} \tag{5.11}$$

where α_{th} representing a maximum given threshold angle. To ensure that the checked point is close to the surface, its height above the reference plane should be limited with the step size. As our hole-filling strategy we are using the ideas presented in [WO03], with the added automation that we fill every hole that is smaller than a given radius. Figure 5.9 presents the results of the proposed resampling algorithm with hole filling for a part of the kitchen dataset.

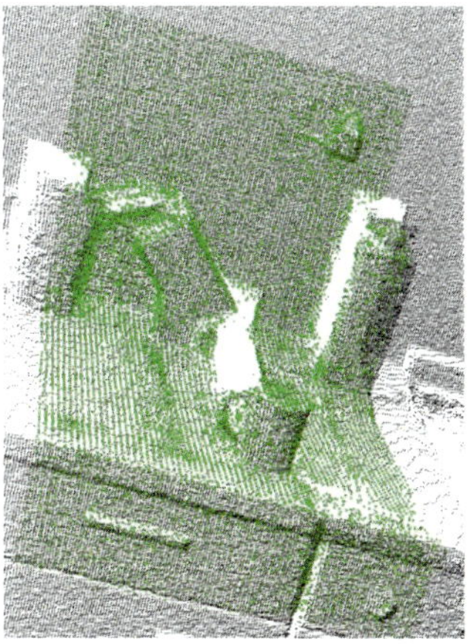

Fig. 5.9 A resampling with hole filling example for a part of the kitchen dataset. The points in $\mathcal{P}$ are shown with black, while $\mathcal{Q}$ is depicted in green.

These points will be projected onto a local reference plane fitted through their k nearest neighbors $\mathcal{P}^k$ from the cloud, thus bringing them in the proximity of the surface. Then, each point $\boldsymbol{q}$ from the projected initial guess $\mathcal{Q}$ is fitted to the surface which approximates the original data using a bivariate polynomial height function in a local Darboux frame. The frame consists of vectors $\vec{\mathsf{U}}\vec{\mathsf{V}}\vec{\mathsf{N}}$, with $\vec{\mathsf{U}} \perp \vec{\mathsf{V}} \perp \vec{\mathsf{N}}$, $\vec{\mathsf{V}}$ being parallel to the fitted plane's normal.

Because $\mathcal{P}$ does not necessarily represent a smooth surface and contains measurement and registration errors, in the first fitting step of our method we apply robust plane fitting using sample consensus methods [MRB09] instead of the usual iterative approaches [ABCO$^+$03] in order to preserve sharp edges in the cloud.

In the bivariate polynomial fitting step, for each point $\boldsymbol{q} \in \mathcal{Q}$, a set of weights $w_{i=1...k}$ is computed for the k nearest neighbors of $\boldsymbol{q}$ as:

$$w_i = \exp\left(-\frac{\|\boldsymbol{q} - \boldsymbol{p}_i\|^2}{h}\right) \tag{5.12}$$

where h defines the Gaussian parameter that corresponds to the sample spacing step, and influences the proportion of the weights of neighboring points. If u, v and n are coordinates along the $\vec{\mathsf{U}}$, $\vec{\mathsf{V}}$ and $\vec{\mathsf{N}}$ axes of the Darboux frame, we can define a number of x height functions $f^j_{(u,v)}$, $j = 1...x$ selected as members of bivariate polynomials, whose sum, when multiplied with the coefficients c_j, approximates the surface in the proximity of $\boldsymbol{q}$'s neighbors as:

$$n_{(u,v)} = \sum_{j=1}^{x} c_j \cdot f^j_{(u,v)} \tag{5.13}$$

To compute the vector of coefficients, we minimize the error function as in [ABCO$^+$03, WO03]. Since the reference plane was computed robustly, the height function $n_{(u,v)}$ defined on it preserves the edges sharper than in regular MLS [RBM$^+$07]. After resampling, surface normals have to be re-computed as the normals of the fitted polynomial at each point. Usually 2nd order polynomials are good enough to approximate the surface because most of the areas are either planar or curved in only one direction. Higher order bivariate polynomials can be fitted where necessary.

Figure 5.10 presents the resampling results for a noisy dataset representing object clusters located on horizontal table planes. Notice that due to the increased noise

Fig. 5.10 A resampling example (right) for noisy point clusters (left) representing objects located on horizontal table planes

present in the raw data (left part), it is almost impossible to recognize the surface of the small jar in the middle of the dataset. The resampled dataset shown on the right side exhibits better geometric properties.

A complete noise filtering and resampling process for the mug presented in Figure 5.8 is shown in Figure 5.11. The results in the left part present the reconstructed mug surface and the superimposed resampled $\mathcal{Q}$ dataset. From middle to right, the remaining images present: i) the raw point cloud data (top) and its erroneous surface reconstruction (bottom); ii) the filtered dataset using the statistical analysis presented in Section 4.2 (top) and the respective surface reconstruction (bottom); and finally iii) the resampled point cloud data with a fixed density (top) and the final surface reconstruction results (bottom).

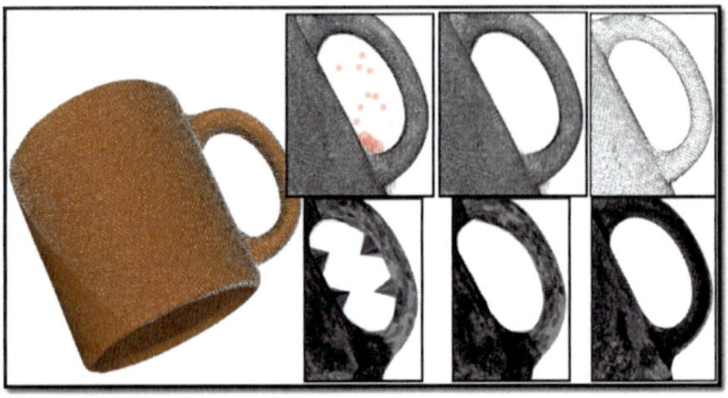

Fig. 5.11 Surface reconstruction examples for a raw (middle), filtered (middle right), and resampled (right) dataset representing a mug

5.3 Related Work

Since the input data leading to the creation of the object map has to be acquired in stages, an algorithm for automatically aligning (registering) the separate views into the same model is needed. One of the most popular registration methods to date is the Iterative Closest Point (ICP) algorithm [BM92, Zha]. The original version uses pairs of nearest 3D points in the source and model set as correspondences, and assumes that every point has a corresponding match. Obviously this is rarely the case with most applications, so a simple distance threshold can be used to discard correspondences in order to deal with partially overlapping scenes. Additionally, to further improve the quality of correspondences, a lot of efforts have been made into the area of feature selection [GMGP05, SLW02], as well as including extra information such as colors [JK97] or normals [BL04] that could improve the correspondence problem. Since ICP requires an exhaustive search through the correspondence space, several variants that address the problem of its computational complexity have been proposed [RL01, And07]. Furthermore, methods for generating initial approximate alignments [GMGP05, SLW02, MPD06], as well as choosing alternate optimization methods [GA05, Fit01], improved the results of the registration process. [NWL$^+$05] presents a system for 3D laser scan registration using ICP based on semantic information (*e.g.* walls, floors, ceilings), which decreases the computation time by up to 30%.

A system able to compute a rough initial guess for the ICP algorithm, using Extended Gaussian Images (EGI) and the rotational Fourier transform is presented in [MPD06]. Extended Gaussian Images are useful for representing the shapes of surfaces and can be approximated using the spherical histogram of surface orientations. While this can provide good transformations which might align two datasets closely for watertight objects, it does not produce satisfactory results for the environment models presented in this book, as the normals alone do not provide enough information about the geometry of the cloud. Our work is based on the idea that relationships between persistent PFH representations in a scene can lead to potentially better alignment candidates.

Once a complete PCD model is obtained, further segmentation and geometrical reasoning algorithms need to be applied, to extract useful information from it. MLS (Moving Least Squares) is a widely used technique for approximating scattered data using smooth functions. [ABCO$^+$03] describes a method to fit polynomials to a point cloud based only on coordinate data and estimated surface normals. To counteract the effects of noise, our approach uses sample consensus like methods and other robust estimators in order to calculate the best reference plane before fitting a high order polynomial [RBM$^+$07]. Another Robust Moving Least Squares algorithm is described in [FCOS05], which differs from ours in that it uses LMedS instead of our sample consensus (SAC) based model fitting approach. To obtain a polygonal model of the world, further surface reconstruction methods can be applied such as [HDD$^+$92, GKS00].

5.4 Summary

This chapter presented two coarse registration methods using Point Feature Histogram representations that can bring datasets acquired from different arbitrary poses into the same coordinate system, creating a consistent global point cloud model. To smooth out the overlapping areas in the resultant point cloud, but also to remove noise and fill holes in the data, we presented a Robust Moving Least Squares algorithm that can resample data and preserve fine details even for noisy datasets such as the ones acquired using our sensing devices.

Chapter 6
Clustering and Segmentation

"Nothing is particularly hard if you divide it into small jobs."

HENRY FORD (1863 - 1947)

Storing and processing large point cloud datasets represents one of the main bottlenecks of a 3D perception system. Though algorithms that can process individual data frames with low computational resources can be written, their capabilities are hindered as larger quantities of data accumulate. A simple example in this direction can be given by the problem of estimating the best planar model that represents a set of points $\mathcal{P}$. Without going into the mathematical details, it suffices to say that the best model can be estimated by computing the distances of all the points $\boldsymbol{p}_i \in \mathcal{P}$ to it, and comparing them with the values obtained from other planar models. It therefore becomes obvious that a smaller dataset $\mathcal{P}_1$ having N_1 points would be processed faster than another dataset $\mathcal{P}_2$ with N_2 points, where $N_2 \gg N_1$.

The key idea then is to create and use mechanisms that would allow big quantities of data to be split up into chunks for faster processing, without essential changes to the expected results, whatever these might be. These resultant data chunks would have a twofold role: i) they would enable data to be interpreted faster; and ii) they would allow the grouping of elements (*e.g.* points) with the same properties (*e.g.* curvature, color) together. The latter has a significant importance for the development of 3D processing algorithms, as in most situations it greatly simplifies their structure and the overall number of steps or iterations that they normally need.

This chapter presents innovative solutions that can be used for the processing of larger 3D point cloud datasets, in the context of point *clustering* and *segmentation*. The concepts are somewhat similar and in some situations they can be used interchangeably. In the context of the work presented here, their main purpose is to group similar structures together, in order to lower the computational resources needed by other subsequent algorithmic steps. In addition, through *segmentation*, we also tackle the problem of fitting a simplified model to some data, such that the model's properties approximate the original data within some error bounds, and provide substantial advantages in terms of the amount of information stored. With a few exceptions only, all the problems tackled in this chapter will be formulated in the context of the kitchen dataset presented in Figure 6.1.

R.B. Rusu: *Semantic 3D Object Maps for Everyday Robot Manipulation*, STAR 85, pp. 75–85.
DOI: 10.1007/978-3-642-35479-3_6

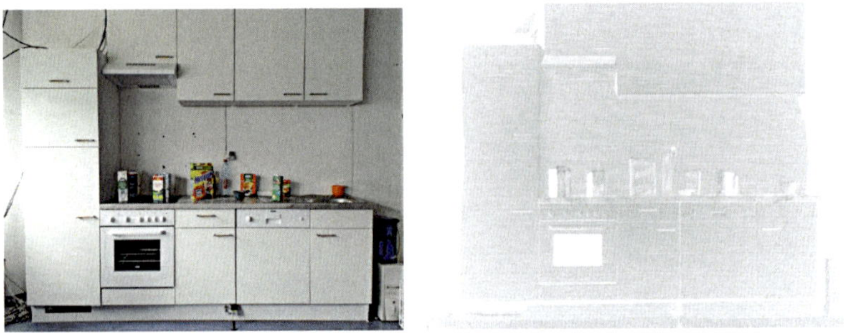

Fig. 6.1 Left: a kitchen scene, right: the resultant point cloud dataset representing the kitchen

6.1 Fitting Simplified Geometric Models

A flavor of data segmentation in itself, the search of certain structures or patterns in a point cloud dataset provides results with an expected quality dependent on two parameters: i) the complexity of the expected structure that is to be searched for; and ii) the noise levels present in the data. Both parameters influence the computational properties of the search. To save time, these model fitting algorithms are embedded with heuristic hypotheses generators such as RANSAC [FB81], which provide upper bounds on the number of iterations that need to be run to achieve a certain required probability of success.

In an ideal world, there would be databases of simplified models (*e.g.* CAD) for all objects present in the world, which would therefore lead to a replacement of most of the points in $\mathcal{P}$ sampled on objects with their underlying models. Supposing that the algorithms that would search and fit these models would work very well, the problem of segmenting surfaces and simplifying the acquired datasets would be solved. Unfortunately, this is rarely the case, and therefore one of the most popular techniques of data simplification is to replace points with 3D geometric primitives. In this context, we refer of a geometric primitive to any basic surface model that is easy to compute, and can approximate the sensed data well enough within some error bounds. Examples include planes, cylinders, spheres, cones, tori, but also higher order bivariate polynomials, or splines, etc.

The reasons why this segmentation and simplification of data has an extreme importance can be better understood in the context of the final goals of an application. Say for example there is a mobile robot that is to move around the world and check for collisions between its body parts and the surrounding environment. If the environment is represented with points or triangular meshes, it would need to estimate a lot of distances from all its body parts to all the points/triangles in the world so that it can compute whether its state is in collision or not. Replacing the points belonging to flat surfaces with large 2D bounding polygons would reduce the number of such computations significantly, as instead of estimating the distances to all the points belonging to that surface, the system would just need to estimate one distance, that to the plane of the polygon.

Another simple application example concerns the problem of grasping objects with a mobile manipulation platform. Say the robot is to pick up a glass or a can of spray (or any cylindrical object for that matter). Estimating the correct grasping points or strategies from a noisy dataset would render the complexity of this problem higher than needed, as the forces between the gripper/hand of the robot and the noisy triangulated surface mesh representing the object would need to be computed all over the surface. Replacing the object with a cylinder instead would simplify the problem, because all grasps along the cylindrical surface could be performed similarly and thus do not need to be computed.

In the context of the applications presented in this book, we are interested in segmenting out such surfaces, with a big emphasis on planar structures which represent a big part of the points in $\mathcal{P}$. To illustrate a few such segmentation examples, we will make use of the kitchen dataset presented in Figure 6.1.

The first problem that is to be solved is to find the area of the kitchen dataset where objects could be placed for pick and place manipulation operations. Making use of contextual knowledge, the system knows that planar horizontal areas (*i.e.* the kitchen counter) can support objects, and will proceed in searching for them. To speed up the search, the algorithm will make use of a Random Sample Consensus (RANSAC) method to generate model hypotheses.

Since the model to be found represents a plane, and since three unique non-collinear points define a plane, the algorithm uses the following steps:

1. Randomly select three non-collinear unique points $\{\boldsymbol{p}_i, \boldsymbol{p}_j, \boldsymbol{p}_k\}$ from $\mathcal{P}$;
2. Compute the model coefficients from the three points $(ax + by + cz + d = 0)$;
3. Compute the distances from all $\boldsymbol{p} \in \mathcal{P}$ to the plane model (a, b, c, d);
4. Count the number of points $\boldsymbol{p}^* \in \mathcal{P}$ whose distance d to the plane model falls between $0 \leq |d| \leq |d_t|$, where d_t represents a user specified threshold.

The last step represents one of the many ways of "scoring" a specific model. Every set of points $\boldsymbol{p}^*$ is stored, and the above steps are repeated for k iterations, where k could be estimated using Equation B.1 for example. After the algorithm is terminated, the set with the largest number of points (inliers) is selected as the support for the best planar model found. From all $\boldsymbol{p}^* \in \mathcal{P}$, the planar model coefficients are estimated in a least-squares formulation, and a bounding 2D polygon (*e.g.* 2D convex hull) can be estimated as the best approximation given the input data of the supporting horizontal plane. Figure 6.2 presents the set of $\boldsymbol{p}^* \in \mathcal{P}$ and the resultant polygon determined as the representation of the kitchen counter from the dataset presented in Figure 6.1.

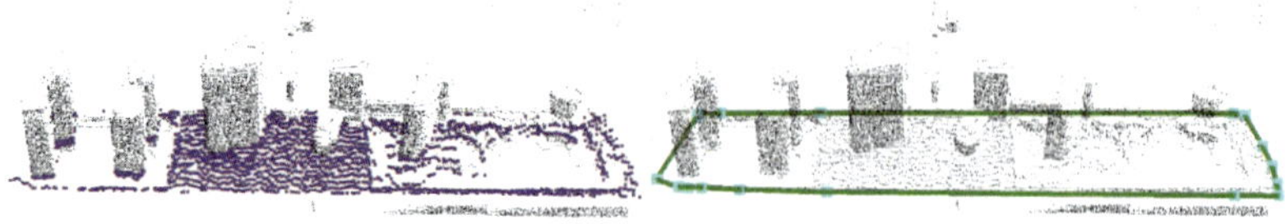

Fig. 6.2 Left: the resultant inliers $p^* \in \mathcal{P}$ that support the best planar model, right: the bounding convex 2D polygon representing the kitchen counter

The above presented computational steps have been simplified to illustrate the basic theoretical process required to fit the geometric plane model. In reality, if we were to find the horizontal support from $\mathcal{P}$, we would need to impose additional constraints, such as the normal of the plane having to be parallel to some global world $\vec{z}$ axis which gives the "horizontal" orientation. A complete and more comprehensive example in given in Chapter 7.

6.2 Basic Clustering Techniques

According to the goals presented earlier, a clustering method needs to divide an unorganized point cloud model $\mathcal{P}$ into smaller parts so that the overall processing time for $\mathcal{P}$ is significantly reduced. Most of the simpler methods in this category rely on spatial decomposition techniques that find subdivisions and boundaries to allow the data to be grouped together based on a given measure of "proximity". This measure is usually represented as a Minkowski norm, with the most popular instantiations being the Manhattan (L1) and Euclidean (L2) distance metrics.

A simple data clustering approach in an Euclidean sense can be implemented by making use of a 3D grid subdivision of the space using fixed width boxes, or more generally, an octree data structure. This particular representation is very fast to build and is useful for situations where either a volumetric representation of the occupied space is needed, or the data in each resultant 3D box (or octree leaf) can be approximated with a different structure.

The 3D grid method however makes sense only for applications requiring equal spatial subdivisions. For situations where clusters can have different sizes, we need a more complex algorithm. To better explain its theoretical steps, let us assume the following application example. Given some input dataset $\mathcal{P}$ as shown in the right part of Figure 6.1 representing an indoor kitchen environment, and a geometric model representing the supporting plane (*i.e.* kitchen counter) that can hold objects for pick and place manipulation tasks (see Figure 6.2), find and segment the individual object point clusters lying on the plane.

To achieve this goal, the system needs to understand what is an *object point cluster* first and what differentiates it from another point cluster. In a more mathematical sense, a cluster is defined as follows. Let $O_i = \{\boldsymbol{p}_i \in \mathcal{P}\}$ be a distinct point cluster from $O_j = \{\boldsymbol{p}_j \in \mathcal{P}\}$ if:

$$\min \|\boldsymbol{p}_i - \boldsymbol{p}_j\|_2 \geq d_{th} \tag{6.1}$$

where d_{th} is a maximum imposed distance threshold. The above equation states that if the minimum distance between a set of points $\boldsymbol{p}_i \in \mathcal{P}$ and another set $\boldsymbol{p}_j \in \mathcal{P}$ is larger than a given distance value, then the points in $\boldsymbol{p}_i$ are set to belong to a point cluster O_i and the ones in $\boldsymbol{p}_j$ to another distinct point cluster O_j. From an implementation point of view, it is therefore important to have a notion on how this minimal distance between the two sets can be estimated. A solution is to make use of approximate nearest neighbors queries via kd-tree representations as explained in Section 4.1. Therefore the algorithmic steps that need to be taken could look as follows:

1. Create a kd-tree representation for the input point cloud dataset $\mathcal{P}$;
2. Set up an empty list of clusters C, and a queue of the points that need to be checked Q;
3. Then for every point $\boldsymbol{p}_i \in \mathcal{P}$, perform the following steps:
   - add $\boldsymbol{p}_i$ to the current queue Q;
   - for every point $\boldsymbol{p}_i \in Q$ do:
     - search for the set $\mathcal{P}_i^k$ of point neighbors of $\boldsymbol{p}_i$ in a sphere with radius $r < d_{th}$;
     - for every neighbor $\boldsymbol{p}_i^k \in \mathcal{P}_i^k$, check if the point has already been processed, and if not add it to Q;
   - when the list of all points in Q has been processed, add Q to the list of clusters C, and reset Q to an empty list
4. The algorithm terminates when all points $\boldsymbol{p}_i \in \mathcal{P}$ have been processed and are now part of the list of point clusters C.

Applied on the remaining point cloud dataset supported by the planar model from Figure 6.2, the proposed algorithm constructs a set of separated Euclidean object clusters as seen in Figure 6.3.

Fig. 6.3 Euclidean clustering results for the remaining points supported by the horizontal planar model presented in Figure 6.2. Each separate cluster is represented with a different color.

6.3 Finding Edges in 3D Data

Together with other feature representations for 3D points as presented in Chapter 4, surface curvature estimates have an interesting characteristic that makes them suitable for data segmentation purposes. Because they approximate the geometry of the underlying sampled surface around a point $\boldsymbol{p}$, they are useful for determining those points with extremely high curvature values which represent the geometric *edges* of the point cloud dataset $\mathcal{P}$. While the concept of an edge in 2D image processing is defined using the gradient of the image, in 3D we can define them as places where the geometry of the scene changes abruptly or sharply.

The thresholding of points having high curvature values can be performed by analyzing the distribution of curvatures in a given point cloud $\mathcal{P}$. The surface curvatures can be estimated by performing an eigenanalysis on the covariance matrix of a patch centered around a query point $\boldsymbol{p}_q$ using Equation 4.15. A graphical representation of the curvatures estimated at each point for the dataset in Figure 6.1, is shown in the left part of Figure 6.4. The right part of the figure presents the distribution of curvatures as a histogram plot. A point can be considered as part of an "edge" either based on a fixed given threshold or by imposing a statistical operator on the previously presented distribution of values (*e.g.* trimean).

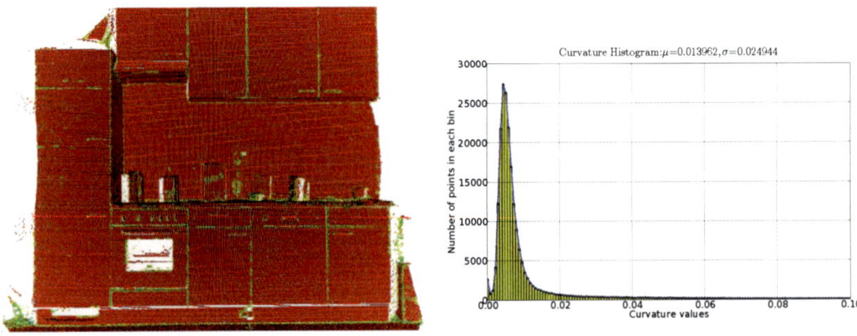

Fig. 6.4 Left: estimated surface curvatures for the kitchen scene from Figure 6.1; right: a plot of the distribution of curvatures

In addition to points with high surface curvature estimates, points located on the geometric boundary of a surface should also appear as edges. The definition of a *boundary point* can be given by analyzing the angles it forms with respect to its neighbors. Let $\Theta = \{\theta_1 \cdots \theta_n\}$ be the list of sorted angles between the lines formed by two points $\boldsymbol{p}_{k_1}$ and $\boldsymbol{p}_{k_2}$ in the neighborhood $\mathcal{P}^k$ with the query point $\boldsymbol{p}_q$. Then $\boldsymbol{p}_q$ is a point located on the boundary of a surface, if:

$$\max(\alpha = \theta_{i+1} - \theta_i) \geq \alpha_{th} \tag{6.2}$$

where α_{th} is a maximum given threshold angle, with a value of $\alpha_{th} = \frac{\pi}{2}$ giving good results in most cases. Figure 6.5 presents the resultant 3D edges of two point cloud datasets acquired in the same kitchen environment.

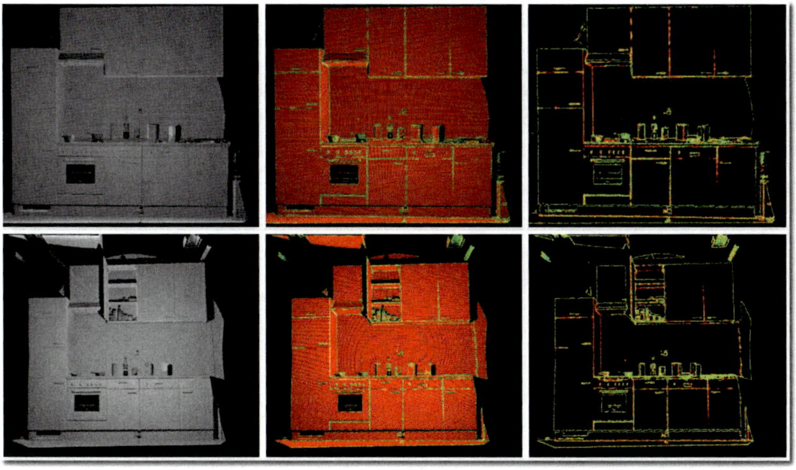

Fig. 6.5 Example of resultant edge representations for two different datasets. Left: the original point cloud datasets, middle: the estimated surface curvatures with low curvatures represented with the red color, and finally right: the points representing the geometric 3D edges.

6.4 Segmentation via Region Growing

The natural extension of the Euclidean clustering algorithm proposed in Section 6.2 for the problem of segmenting points with similar properties together, is to include additional information in the checks performed at step 3, such as the point's color, or an information regarding its surrounding geometry, etc.

An example would be to find and segment connected regions in $\mathcal{P}$ with smooth surfaces that are separated by edges, *i.e.* sudden changes in curvature and normal orientation. The algorithm would start by adding a point $\boldsymbol{p} \in \mathcal{P}$ with a low curvature value to a queue (initially empty), and verify each of its $\boldsymbol{p}_k$ neighbors to see whether it could belong to the same region as $\boldsymbol{p}$. The criteria based on which a neighbor $\boldsymbol{p}_k$ is said to belong to the same region is:

$$\arccos(\langle \vec{\boldsymbol{n}}, \vec{\boldsymbol{n}_k} \rangle) \leq \alpha_{th}, \tag{6.3}$$

where $\vec{\boldsymbol{n}}$ and $\vec{\boldsymbol{n}_k}$ are the surface normals at the points $\boldsymbol{p}$ and $\boldsymbol{p}_k$. Therefore, if the angle formed by their normals does not exceed a given threshold α_{th}, the point $\boldsymbol{p}_k$ can be added to the current region started from $\boldsymbol{p}$. An additional check can be performed on its estimated surface curvature, which needs to be bound by a predefined σ_{th} curvature. If $\boldsymbol{p}_k$ fulfills the extra check as well, it will be added to the queue list, and the procedure is restarted.

This method is similar to a region growing approach, with the difference that it propagates the growth along directions of lower curvature first. Points will stop being added to the current region if none of their neighbors fulfills the angle criteria and the queue of points is empty (meaning that all points with a small curvature have already been considered). The algorithm can be extended to automatically estimate a good value for σ_{th} as we have shown in our previous work [RMB$^+$08b].

By applying the proposed method to the dataset presented in Figure 6.4, we can obtain solutions such as the ones presented in the left part of Figure 6.6. The right part of the figure presents the resultant remaining point clusters after running the segmentation algorithm, grouped into Euclidean cluster using the method presented in Section 6.2.

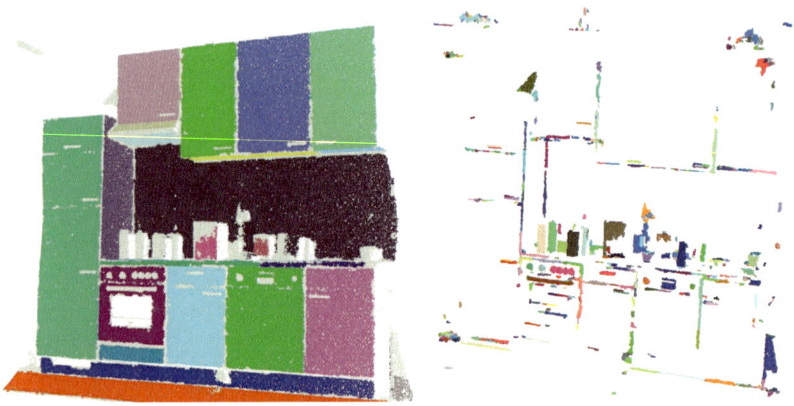

Fig. 6.6 Segmentation results for the kitchen: accepted regions (left) and remaining points (right). Each region cluster is marked with a different color. All remaining points from the right side are marked with gray in the left part of the figure.

Figure 6.7 presents the points located on the boundaries of each segmented region, using the boundary point detection algorithm presented earlier.

Fig. 6.7 Left: the segmented regions from Figure 6.6 together with their respective boundary points; right: the boundary points of each segmented region

6.5 Application Specific Model Fitting

Following the method presented in Section 6.1, the model to be searched for can be replaced with different geometric shapes, depending on the goals of a particular application. For example, once a set of planar regions has been identified in $\mathcal{P}$, they could be replaced with 2D polygons that describe the regions using fewer points, thus leading to a more compact representation of the world.

Additionally, from the set of polygons that replace 2D planar surfaces, a subset could be replaced by 3D cuboids (rectangular parallelepipeds), given the fact that we are working on a very specific application example involving pieces of furniture and kitchen appliances. These cuboids could be selected from the set of candidates which respect constraints with respect to their orientation (vertical and horizontal), and satisfy a certain size criterion. Their final shape could then be obtained by projecting the vertical bounding 2D polygons onto the walls.

In a general case, the bounding 2D polygonal surface of the planar regions from Figure 6.7 could be obtained from a convex hull representation of the plane inliers, similar to the approach presented in Section 6.1. Here however, the system could make use of an even simpler model, and specialize its routines to use that, since the approximations of the planar regions refer to the frontal faces of furniture candidates.

A solution is to model 2D quads from the list of boundary points of each planar region, by fitting the best four lines to it [RMB+08b, RMB+08a]. Pairs of lines close to perpendicular are therefore accepted as candidates if their four intersection points form a 2D quad with reasonable dimensions. Let $\vec{z}$ be a global coordinate axis pointing up, such that the resultant planar faces of the furniture candidates have their normals perpendicular to it. The algorithm presented in Section 6.1 could then be modified as follows:

1. Randomly select two unique points $\boldsymbol{p}_i, \boldsymbol{p}_j \in \mathcal{P}^*$, where $\mathcal{P}^*$ represents the set of points located on the boundary of a segmented planar region;
2. Compute the line model coefficients from the two points $(\boldsymbol{p}_i, \vec{d}_i)$, where $d_i = \boldsymbol{p}_j - \boldsymbol{p}_i$;
3. Check if the line model is parallel or perpendicular to $\vec{z}$, and go back to step 1 if not;
4. Compute the distances from all $\boldsymbol{p} \in \mathcal{P}$ to the line model;
5. Count the number of points $\boldsymbol{p}^* \in \mathcal{P}^*$ whose distance d to the line model falls between $0 \leq |d| \leq |d_t|$, where d_t represents a user specified threshold.

Once the four perpendicular lines representing the frontal face are obtained, we can take the closest wall parallel to the 2D quad formed, and simply project the quad along its plane normal to the wall. Figure 6.8 depicts the resultant 2D quads represented by the set of lines fitted to the boundary points and their four intersection points on the left part, and the projected 3D rectangular parallelepipeds representing the kitchen furniture containers on the right.

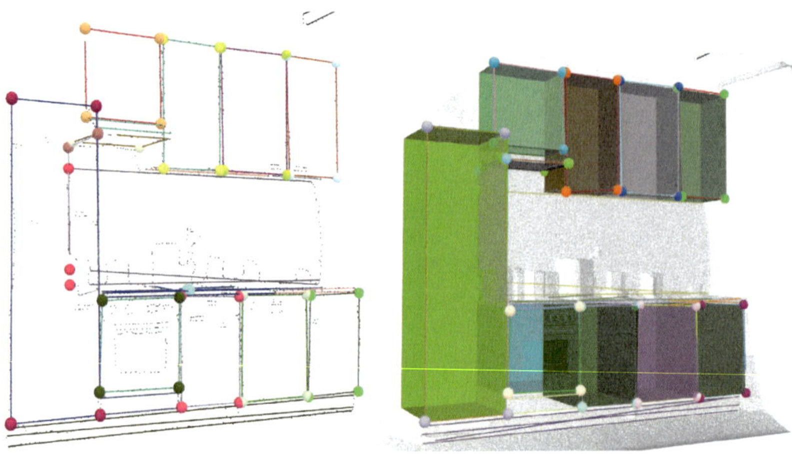

Fig. 6.8 Furniture candidates representations using 2D quads (left) and 3D cuboids (right) projected on the closest parallel wall

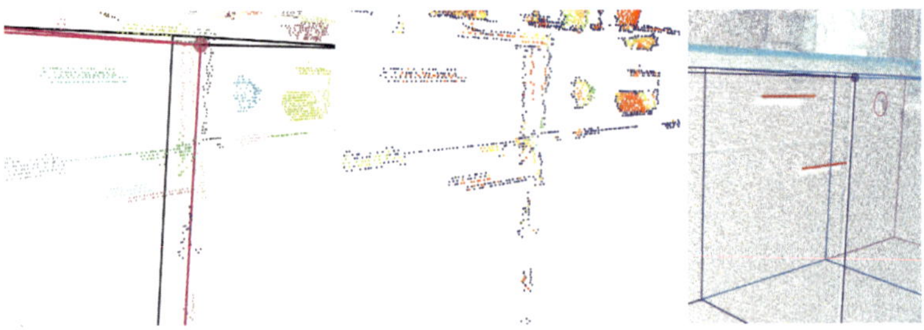

Fig. 6.9 The remaining clusters of points projected on their associated cuboids (left); Estimated boundary points of each cluster (middle); Handles and knobs segmentation via line and circle fitting (right)

Another specific model fitting application can be tackled in the context of finding the fixtures (handles and knobs) of the previously detected furniture candidates, such that they could be used by a mobile manipulation platform in its tasks to open and close them. Looking at the right part of Figure 6.6, we can observe that many of the remaining point clusters represent the set of handles and knobs located on the frontal doors of the cupboards, the oven, etc.

For finding out their exact location, we need to look at clusters of points which did not get assigned to one of the segmented planar regions, but whose set of neighbors are inside a planar region. These can then be selected for an analysis if they are at a distance d_i from the surface, in between some predefined thresholds:

$$d_{\min} \le d_i \le d_{\max} \tag{6.4}$$

where a fixture (*e.g.* handle) could be located and operated by the robot's end effectors.

An implementation example would be to project the candidate cluster representing a possible fixture onto the door plane of the respective region, and extract its boundary points. Since the application is limited to the segmentation of handles and knobs, these can be approximated with lines and circles in a two-dimensional space, as seen in Figure 6.9.

6.6 Summary

This chapter presented a set of mechanisms that allow big quantities of data to be split into chunks for faster processing, through the use of clustering and segmentation techniques. Besides lowering the computational constraints of processing algorithms, the individual data chunks allow the grouping of points with the same geometric or texture properties together, which leads to significant simplifications of the resultant models.

Part II

Mapping of Indoor Environments

Chapter 7
Static Scene Interpretation

> *"All science is static in the sense that it describes the unchanging aspects of things."*
>
> FRANK KNIGHT (1885 - 1972)

This chapter reports on experiences regarding the acquisition of hybrid Semantic 3D Object Maps for indoor household environments, in particular kitchens, out of sensed 3D point cloud data. The proposed approach includes a comprehensive pipeline, with geometric mapping and learning mechanisms, for processing large input datasets and for extracting relevant objects useful for a personal robotic assistant to perform complex manipulation tasks. The type of objects modeled are those which perform utilitarian functions in the environment such as kitchen appliances, cupboards, tables, and drawers. The resultant model is accurate enough to use it in physics-based simulations, where doors of 3D containers can be opened based on their hinge position. The resultant map is represented as a hybrid concept and is comprised of hierarchically classified objects and triangular meshes used for collision avoidance in manipulation routines.

The proposed Semantic 3D Object Maps interpretations are motivated by the fact that autonomous personal robots performing everyday manipulation tasks such as setting the table and cleaning up in human living environments must know the objects in their environments: the cupboards, tables, drawers, the fridge, the dishwasher, the oven, and so on. The knowledge about these objects must include detailed information about the objects geometry, and structural knowledge as: a cupboard consists of a container box, a door with hinges and a handle. Additionally, there is a need for acquiring functional knowledge that enables the robot to infer from the position of a handle on a door the side to which the door opens.

To this end, the computational problem can be formulated as follows: given a 3D point cloud model of an environment as depicted in the left part of Figure 7.1, segment the point cloud into subsegments that correspond to relevant objects and label the segments with the respective category label (see Figure 7.1 right).

The resultant labeled object model is meant to represent the environment as best as possible given the geometry present in the input data, but its accuracy does not have to be absolute with respect to the true world model. Instead, the object model is

R.B. Rusu: *Semantic 3D Object Maps for Everyday Robot Manipulation*, STAR 85, pp. 89–108.
DOI: 10.1007/978-3-642-35479-3_7

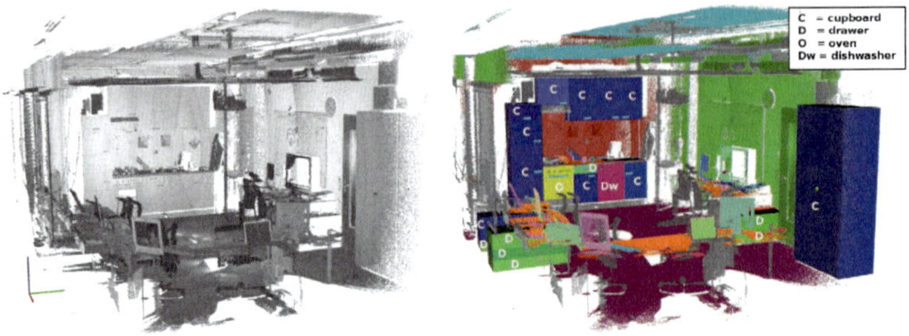

Fig. 7.1 Left: a snapshot of the kitchen dataset: 16 registered scans shown in intensity (grayscale), comprising roughly 15 millions of points. The world coordinate system depicted on the bottom left shows $\vec{x}$ with the red color, $\vec{y}$ with green, and $\vec{z}$ with blue. Right: a Semantic 3D Object Map representation of the same kitchen environment. The representative planar areas are shown in different colors (tables - orange, floor - dark purple, walls - green and red, ceiling - cyan), and 3D cuboid containers are marked with their appropriated labels (cupboard, drawer, oven, etc). The remaining unclassified points are shown in gray.

considered as an intermediate representation that provides candidate objects which are to be validated through subsequent processing steps. These steps include vision based object recognition, active exploration like for example opening the drawers and doors that were suggested, and classifications based on the role that an object has in a certain activity (*i.e.* activity recognition). For this reason, the main objective of the proposed mapping system is to compute the model as quickly as possible using solely the geometric information contained in the point cloud, and have results that approximate the true world model.

The term *hybrid* mapping refers to the combination of different data structures in the map, such as: points, triangle meshes, geometric shape coefficients, and 2D polygons. Different tasks require different data structures from this map. For example, 3D collision detection usually requires either a triangle mesh representation or a voxelization of the underlying surface, while object classification might use the geometric shape coefficients. The hybrid Semantic 3D Object Map in the proposed implementation is therefore comprised of two different types of maps:

- a Static Semantic Map containing the relevant parts of the environment including walls, floor, ceiling, and all the objects which have utilitarian functions in the environment, such as fixed kitchen appliances, cupboards, tables, and shelves, which have a very low probability of having their position in the environment changed (see the right part of Figure 7.1);
- a Triangulated Surface Map, used for 3D path planning, and collision avoidance for navigation and manipulation, using the techniques presented in [MRB09].

The former can be created in two different ways: either by using a set of heuristic rules, or by learning classification models that can label the segmented regions with the appropriate object class.

7.1 Heuristic Rule-Based Functional Reasoning

The Static Functional Map includes semantically annotated parts of the environment, which provide useful information for a mobile personal assistant robot in fulfilling its tasks. These parts are thought of as being *unmovable* or *static*, that, is with a very low probability of having their position changed in the environment (though this is not a hard constraint for the system presented herein, in the sense that model updates are possible). A separation of the objects based on their functions is given as follows:

- box-like containers which can contain other objects and have states such as open and closed: cupboards, drawers, and kitchen appliances;
- structural components of the environment: walls, floor, ceiling, and doors;
- supporting planar areas: tables, tops of sideboards, shelves, counters, etc.

The first category is comprised of kitchen appliances, cupboards, and drawers – all thought of as being central actors for a robot task, as they contain objects of interest that can be manipulated. In the second category we place those parts of the environment which support objects in the first category and also help the robot navigate around, like the walls in the room, the floor and the ceiling. Finally, the third category is comprised of horizontal (with respect to a fixed world $\vec{z}$ axis) planes on which objects that can be manipulated in pick and place scenarios are located.

The left part of Figure 7.1 presents a 360° view, comprised of 16 registered scans of the kitchen lab used for the experiments presented in this chapter. Note that the attached coordinate system presented on the left side of the figure is shown for

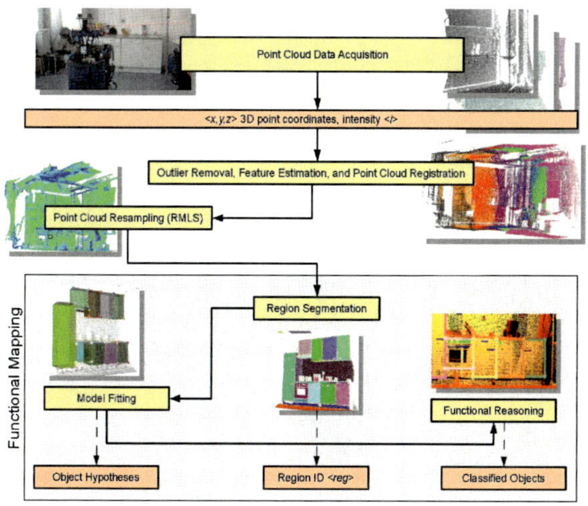

Fig. 7.2 An overview of the proposed Semantic 3D Object Mapping system architecture

orientation purposes only, that is to depict the general $\vec{x}\vec{y}\vec{z}$ directions ($\vec{x}$ - red, $\vec{y}$ - green, $\vec{z}$ - blue).

To obtain Semantic 3D Object Maps from individual raw scans, a complete and compact representation of the world has to be constructed first, that is, the separate views have to be registered together and transformed into a global coordinate framework. Since the topic of this chapter focuses more on the functional mapping aspects (see the proposed system architecture in Figure 7.2), the modules which fall outside its scope will not be re-addressed here, as they have already been covered in Chapters 4, and 5. After a complete point cloud model has been obtained, it will go through the following processing steps:

- a region segmentation step for extracting planar regions, based on estimated surface normals and curvatures (see Section 6.4);
- a model fitting step for identifying object candidates (cupboards, tables, etc.), represent them using cuboids, and search for connected handles and knobs (see Section 6.5);
- a functional reasoning step where object candidates are verified and their functionality is deduced from the geometry data.

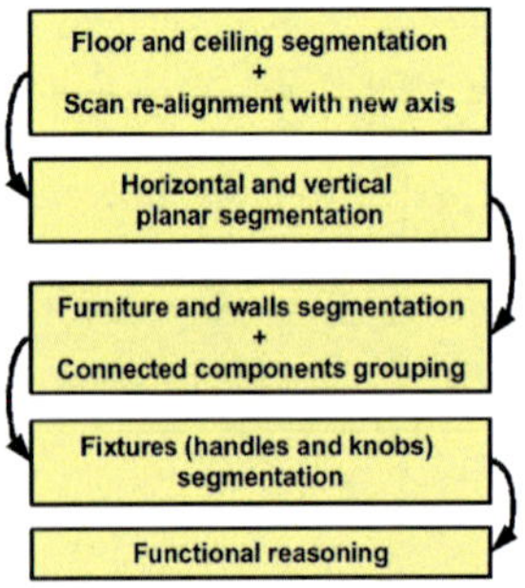

Fig. 7.3 Static Functional Map processing pipeline

Figure 7.3 presents the individual steps in the processing pipeline that leads to the creation of the Static Functional Map. The coordinate system of each individual scan (before registration) is identical to the local coordinate system of the robot from the data acquisition position. By assuming that the $\vec{z}$ axis of this local coordinate system remains constant (pointing upwards) throughout the acquisition of all datasets, the mapping system performs a search for the bottommost and topmost planar areas in each scan. Once found, these areas will constitute the floor and the ceiling of the room in that scan, and will be marked appropriately in the cloud. The search constraints are simple commonsense heuristic rules that impose that the floor should be roughly at a distance d_f on $\vec{z}$ from the origin of the robot coordinate system, equal with the robot height, while the ceiling should be higher than a given threshold. From an implementation point of view, d_c can be chosen as $d_c \geq (2.5m - d_f)$. To achieve near real-time performance, every dataset is downsampled using a fixed width octree structure and a search for planar areas respecting the above mentioned constraints is performed in a sample consensus formulation. Once determined, the plane equations will be refined by including the original points in the scan and refitting the model. After the floor has been identified and labeled as such, the coordinate system of the dataset will be snapped to the floor by imposing that its $\vec{z}$ axis should be perfectly perpendicular to the floor.

After performing global registration on all scans, the coordinate system of the resultant dataset might be changed. However, a straightforward transformation can be estimated between any of the individual scans and the 360° one, and thus the $\vec{z}$ axis of the 360° dataset can be re-aligned with the $\vec{z}$ axis of the robot to simplify subsequent processing steps.

The next step in the mapping pipeline is to detect large planar areas perpendicular to the floor and the ceiling which could represent the walls of the room. The assumption is that for the initial Static Functional Map (at its creation), scans can be acquired in such a way that large vertical portions of the room are covered, and therefore it is most likely that at least one wall is observed in each scan. This means that the system does not need to process multiple scans or the 360° dataset to get the walls of the room (though that is possible as well), but rather incrementally detect and grow them as new scans come in.

To detect these areas efficiently, we first look into the Extended Gaussian Image (EGI) of the point cloud (see Figure 7.4) and take all the estimated point normals which are perpendicular to the $\vec{z}$ axis. Then a sample consensus search is performed in the normal space in order to obtain the other two principal directions $\vec{x}$ and $\vec{y}$. If only one direction is found, the other will be computed directly using a cross product operation. Once the three major axis are determined, the coordinate system of the dataset will be re-aligned to match the newly determined $\vec{xyz}$ coordinate system.

Having the principal directions of the scan aligned with a defined $\vec{xyz}$ coordinate system, makes searching for planar regions highly efficient, as only points whose normals are approximately perpendicular to the plane we try to determine will be considered. Therefore, the search on all three major directions is highly parallelized in order to discover the planar supports for walls, cupboard doors, tables, etc. Figure 7.5 presents the major planar decomposition results for the input kitchen

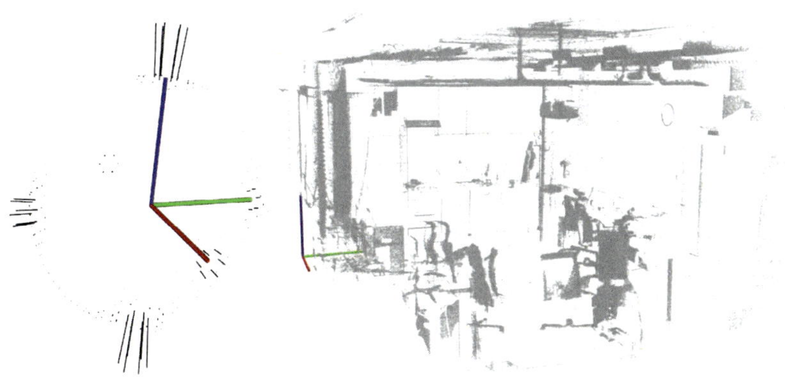

Fig. 7.4 Left: the Extended Gaussian Image (EGI) of the point cloud dataset. As seen, most of the estimated point normals are found around the principal $\vec{xyz}$ directions, accounting for approximately 85% of the entire dataset. Right: the remaining points after major planar area segmentation.

Fig. 7.5 Major planar area segmentation in the kitchen dataset. Each area is shown in a different (random) color, while the remaining unsegmented points are shown in gray.

dataset, while the right part of Figure 7.4 presents the remaining points after all segmented planar areas have been removed.

The next step is to verify which of the vertical planar regions could possibly be a wall and separate it from the rest. Since the planar equations of the ceiling and the floor are already known, they can be used to determine the height of a vertical region with respect to them. The goal here is to differentiate between walls and other types of vertical planes which will be considered unanimously as possible furniture candidates. Therefore, the vertical regions which contain points higher than a distance d_{wf} from the floor or lower than a distance d_{wc} from the ceiling, will be marked as walls. For the purpose of the application presented herein, it is not extremely important if these are actual walls or not, as d_{wf} will be selected to be high enough ($\geq 2.5m$) so that those regions are unreachable by the robot anyway. The left part of Figure 7.6 presents the vertical planar candidates which satisfied the above mentioned criteria and were marked as walls. Notice that the regions do not have to be continuous, as all the points which have the same plane equation will be marked as walls (*e.g.* the part of the wall on which the kitchen cupboards are installed has points above the cupboards satisfying the wall criteria, and therefore will be marked as wall too).

After finding and marking all the walls, the remaining vertical planar areas are marked as possible furniture faces. However some of them are too small, and must not be considered at this stage by the system. To prune these regions from the list, their area is first estimated and only the ones which are large enough to sustain the smallest piece of furniture that is to be considered, will be retained. The right part of Figure 7.6 shows the remaining furniture candidates.

Fig. 7.6 Left: planar areas segmented as walls (based on the ceiling closeness criterion); right: planar areas segmented out as possible furniture faces

A further refinement will be done based on whether these candidates have some sort of a fixture (handle or knob) for operating or opening/closing them. This means that any planar area that is a furniture candidate, but does not contain a fixture, cannot be manipulated by the robot in any way, and thus should not be considered either. To do this however, the furniture candidates must first be broken into minimally connected planar regions, and a search for fixtures for each of these regions needs to be performed independently. To do this, the boundary points of each individual furniture candidate are first estimated, followed by a segmentation step using a region growing approach. All under-segmentation cases will be solved through the use of heuristic rules as shown below.

The region segmentation method is similar to the algorithm presented in Section 6.4, with two differences. First of all, the regions are grown until we hit boundary points instead of looking at the estimated surface curvature at each point. This has the advantage that the system does not need to estimate any additional curvature thresholds. Secondly, to speed up region growing, the algorithm makes use of an octree decomposition, that is, if no boundary points are found within an octree leaf, all points are automatically added to the current region. The segmentation results are shown in Figure 7.7.

Fig. 7.7 Left: furniture candidate faces shown with their respective boundary points (marked with red). Right: refined planar furniture candidates after segmentation. Due to under-segmentation, certain candidates were not split correctly (*e.g.* large cabinet on the left of the kitchen).

For each of the segmented furniture faces candidates, a search is performed for points lying in their vicinity, which could contain fixtures such as handles and knobs. The algorithm for extracting fixtures is similar to the one already presented in Section 6.5 and consists of the following steps:

- compute the boundary points of each furniture face candidate;
- obtain the two directions perpendicular to the normal of the planar area, and find the best (*i.e.* highest numbers of inliers) four oriented lines, two in one direction and two in the other direction in a sample consensus formulation;
- get the four points which form the 3D rectangle approximating the planar region;
- get all points in the vicinity of this rectangle at a distance $0cm \leq d_i \leq d_t$ (in our implementation we selected $d_t = 5cm$) and compute their boundary points;
- fit 3D lines and 3D circles to these boundary points in a sample consensus formulation, score the candidates, and select the ones which minimize the Euclidean distance error metric.

To refine the final fit, a non-linear optimization is applied using a Levenberg-Marquardt algorithm. Figure 7.8 presents the segmentation and classification of all handles and knobs found on candidate furniture faces for the input kitchen dataset.

The proposed mapping approach relies on a set of heuristic rules imposed on the entry data. These rules decide whether and how areas of the map will be considered in a certain geometric processing step. To achieve an overall consistency in the results, these rules are strict, in the sense that: if a certain area does not fulfill a

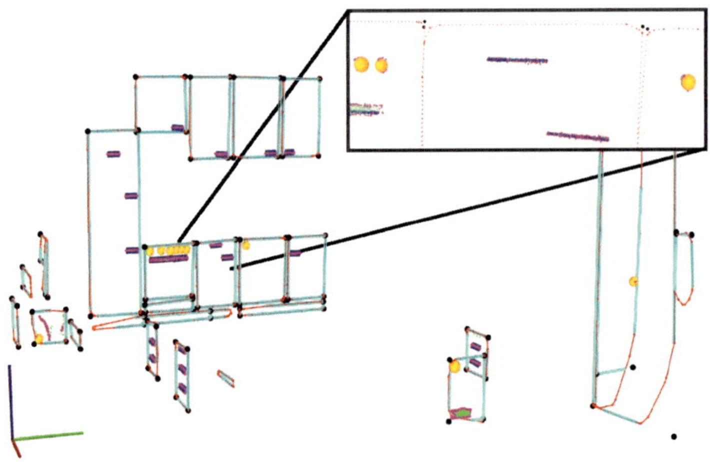

Fig. 7.8 Segmentation and classification of fixtures (handles and knobs) on furniture faces (see Figure 7.7). Handles are drawn with blue lines over their inlier points (in magenta), knobs with orange circles, and each planar area is bounded by 4 corners (in black) and 4 perpendicular lines (in cyan). The red dotted-lines represent the convex boundaries of each point region.

single criterion, the system simply leaves it out, and re-processes it when more data is available.

For example, planar regions with an area smaller than $0.0625sqm$ ($\approx 0.25m \times 0.25m$) are not considered for further interpretation until more data is available in that region. These rules represent sensible thresholds for the segmentation methods presented here, and could be relaxed even further if the quality of the acquired point cloud dataset would be higher (*i.e.* by using a better sensing device).

A different category of rules altogether are represented by the presence or absence of certain *conditions* for a given object candidate. In this respect, the fixtures (handles and knobs) detected as lying near furniture surface candidates are used to decide:

- what class of furniture could the candidate belong to;
- where is the hinge of the door located;
- whether the candidate is a composition of candidates due to under-segmentation.

The latter refers to the presence of multiple handles for a furniture candidate. From an implementation point of view, we impose that one piece of furniture should contain only one handle (drawers and cupboards are treated as individual units and not just as part of a cabinet), and thus multiple handles indicate that under-segmentation occurred in that area. Therefore, the presence of multiple fixtures triggers an internal re-segmentation process for an area, by looking to fit lines that might split the candidate into multiple parts which have not been detected previously because they were either incomplete, or had a low number of inliers.

In detail, the algorithm selects each region which contains multiple handles attached to it, and proceeds to split the region into N smaller regions, where N is the total number of handles found. The region re-segmentation is based on 3D line fitting in a sample consensus formulation in the space of boundary points located inside the region. Since there is no guarantee that there are enough boundary points

Fig. 7.9 Refining furniture candidate faces based on the presence or absence of multiple fixtures

so that a line segment can split the region completely, the resulting segments are grown until they hit the borders of the region. After all lines are found, the points in the region together with the handles are re-associated to the newly created smaller regions. The results before and after this functional segmentation step are presented in Figure 7.9.

After re-segmentation, the location of the door's hinge is inferred as follows: a) if the handle associated with a frontal face of a furniture unit is located near one of the edges of the unit, then the hinge of the door is located on the opposite edge; and b) if however, its position is roughly in the middle of the unit's geometric center, then the unit is most likely a drawer so it doesn't contain a hinge.

Another commonsense assumption that can be made is that a furniture unit with one or several knobs close to each other can in fact be a kitchen appliance, like an oven or a dishwasher. This comes from the simple fact that knobs are usually used to change settings and influence the functioning of a device instead of just opening doors, for which handles are more appropriate. Most kitchen designs depict a kitchen oven as a cupboard-like unit with one large handle for the oven door, and several knobs above it for operating the burners.

By taking all the above rules into consideration, and applying them to the list of detected furniture candidates, a simple hierarchical object model such as the one presented in Figure 7.10 can be created. This hierarchical model will annotate (as best as possible given the input data) different class object labels for each candidate. Therefore, a cuboid-like container with a fixture (handle or knob) present near its frontal face, can be labeled as:

- a cupboard – if it has only one handle or knob, and a fairly big surface area (note that very large cupboards are considered as closets);
- a drawer – if it has one handle located near the geometric center of its frontal face, and a smaller surface area;
- a kitchen appliance – if several knobs and at most one handle are present. A further decomposition into ovens and dishwashers, or washing machines can be done, based on the assumptions mentioned above (*e.g.* an oven's design).

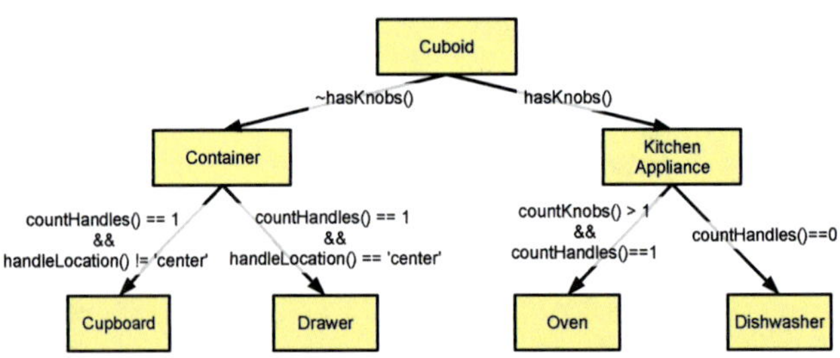

Fig. 7.10 A Hierarchical Object Model scheme using heuristic commonsense rules

The first two labels represent simple containers for smaller types of objects. Please note that kitchen appliances like a refrigerator which are usually built to resemble cupboards in modern kitchen designs cannot be differentiated solely based on their appearance. This however is not a major drawback as a refrigerator is indeed a cupboard-like container, with the difference that only specific types of items are stored inside.

From an implementation point of view, the input datasets are spatially decomposed using fixed-leaf dynamic octrees (see Figure 7.11), to achieve fast computation times for each individual processing step in the proposed mapping pipeline. This allows the system to use different levels of detail, depending on the goals of the application. Each individual new scan that is acquired stores a couple of millions of points, and working directly on it would not only render some of the geometric methods slow, but also increase the necessary memory requirements. The workaround is to build the octree structure as early as possible and then grow it incrementally when new registered data comes in.

Fig. 7.11 Two different levels of detail representing approximately 1% (level 8), respectively 0.2% (level 7) of the total number of points from the original dataset, created using an octree

The room shape assumptions presented at the beginning of the section do not represent hard constraints for the system, and their usage only leads to a significant performance increase in our processing pipeline. Without any generality loss, if the search for principal directions and re-orientation steps would be left out, the algorithms would fall back to the default behavior where planar areas would be searched for within the entire occupied octree cells, leading to the same results with the only drawback being that the entire procedure would just take longer. Also, the system's operation is not hindered in any way if multiple rooms are present in the scans, as it does not assume four bounding walls.

The set of commonsense heuristic rules that the proposed mapping approach bases its functionality on will be extended in the next section through the usage of a comprehensive classification scheme. This way, the system will learn parameters from a set of labeled training data and thus substitute the need for a user defined rule, like for example: a wall needs to have some points higher than a certain distance height d_{wall} from the floor, etc.

7.2 Learning the Scene Structure

The map learning can be approached by designing a multi-layered geometric feature set that can extract meaningful information from training examples and help create a good machine learning classifier model. The model could then be used to generate labeled object hypotheses using just the geometric data contained in the acquired point clouds.

The architecture of the system depicted in Figure 7.12 is similar to the one from the previous chapter as proposed in Figure 7.2. In particular, the Semantic Mapping pipeline includes:

- a highly optimized major planar decomposition step, using multiple levels of detail (LOD);
- a region growing step for splitting the planar components into separate regions;
- a model fitting step for fixture decomposition;
- finally a 2-layers feature extraction and classification step.

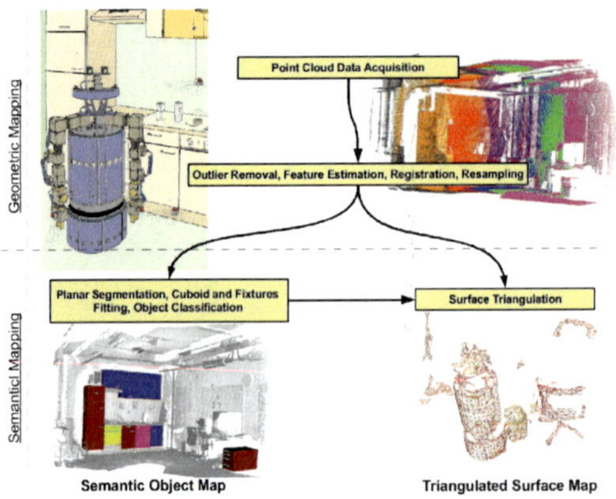

Fig. 7.12 The architecture of the proposed mapping system, and the 2 different types of maps produced. The input data is provided from the laser sensors installed on the robot's arms via the Point Cloud Data Acquisition module, and is processed through a Geometric Mapping pipeline resulting in a *PCD world model*. This model constitutes the input for the separate components of the Semantic Mapping module.

Figure 7.14 describes the 2-layers feature extraction and classification framework employed in the proposed Semantic Mapping system. Instead of learning a single global model, the system makes use of proven geometric techniques for splitting the data into clusters first, and computes separate features and a separate model for each of these clusters. The two defined layer-1 clusters are composed of the horizontal planes, and the vertical planes respectively. By treating them separately, the features that need to be computed are greatly simplified, false positives are removed, and in general the classification results are improved. Additionally, once a set of *furniture faces* labels are obtained from the classifier for vertical planes, the system proceeds at extracting object fixtures (*e.g.* handles and knobs) and estimates a set of layer-2 features which will help at separating furniture object types into drawers, cupboards, kitchen appliances, and vertical side faces respectively. A final advantage of this scheme is that there is no need to estimate all possible features for all planar candidates, but the architecture allows for a flexible scheme, where the segmentation and estimation of features is done *as needed*, starting with simple ones first (*i.e.* horizontal planes). Therefore, the overall system will benefit from a reduced computational complexity.

Since the world coordinate frame is defined with the $\vec{z}$ axis pointing upwards, the planar models to be searched for are divided into two categories:

- *horizontal* planes, *i.e.* those whose normal is parallel with the $\vec{z}$ axis;
- *vertical* planes, *i.e.* those whose normal is perpendicular to the $\vec{z}$ axis.

The first category will include structural components of the environment such as the floor and the ceiling, as well as planar areas which can support movable objects, such as tables, shelves, or counters (see Figure 7.13 left). The second category will devise the walls of the room, and all the faces of the furniture and kitchen appliances in the room (see Figure 7.13 right). For each segmented planar area, the system then estimates its boundary points, and a region growing approach similar to the one presented in Section 7.1 is performed in order to determine the set of planar *regions*. These constitute the input to the feature extraction modules as presented in Tables 7.1, 7.2 and 7.3.

Fig. 7.13 Left: all horizontal planes found in the scene; right: all vertical planes found in the scene

The description of the layer-1 features is given in Tables 7.1 and 7.2. The first set of features will be computed only for horizontal planes. Once a model that can separate horizontal planes into the object classes mentioned in Figure 7.14 is learned, the resultant ceiling and floor object models will be used to generate the layer-1 features for vertical planes.

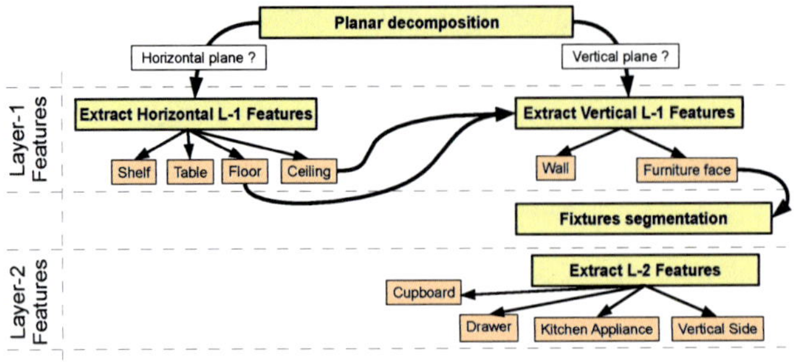

Fig. 7.14 The 2-layered feature extraction and object classification framework used in the proposed mapping system. The light red boxes represent the output classification labels.

Table 7.1 Layer-1 features for horizontal planes

Feature	Notation	Description
Height	$\mathcal{H}_h$	the height of the planar model on $\vec{z}$ with respect to the world coordinate frame
Length	$\mathcal{L}_h$	the length along the first principal component
Width	$\mathcal{W}_h$	the length along the second principal component

The vertical planar classification separates walls from furniture candidates. Since the planar equations of the ceiling and the floor are already known from the horizontal planar classification, they are used to determine the height of the vertical region with respect to them. The goal is to differentiate between walls and other types of vertical planes which will be considered unanimously as possible furniture candidates. Therefore, the vertical regions which contain points close to the ceiling might be classified as walls. For the purposes of the experiments presented herein, it is not extremely important if these are actual walls or not – what matters is that those regions are high enough that they are unreachable by the robot anyway.

Since the feature spaces are relatively simple, the choice of using the right machine learning classifier is greatly simplified.

To differentiate between various possible furniture types, the proposed mapping architecture implements a secondary type of features which are to be computed only for the vertical planar regions classified as furniture candidates. This set of features (see Table 7.3) take into considerations constraints such as the number of handles and knobs present as lying on the planar region, as well as the distance between the center of the fixture and the center of the region.

Following the classification results for object types which employ fixtures towards one of the edges of the planar face supporting them (*e.g.* cupboards), the system will estimate the door opening hinge as being on the opposite edge.

Table 7.2 Layer-1 features for vertical planes

Feature	Notation	Description
Height	$\mathcal{H}_v$	the actual length along the $\vec{z}$ axis (*i.e.* $\lvert \boldsymbol{M}_z - \boldsymbol{m}_z \rvert$ where $\boldsymbol{M}_z$ and $\boldsymbol{m}_z$ are the points with the maximum respectively minimum $\vec{z}$ values)
Floor distance	$\mathcal{D}_v^f$	the distance to the floor model (*i.e.* $\lvert \boldsymbol{m}_z - \boldsymbol{p}_f \rvert_z$ where $\boldsymbol{m}_z$ is the point with the minimum $\vec{z}$ value, and $\boldsymbol{p}_f$ is a point on the floor)
Ceiling distance	$\mathcal{D}_v^c$	the distance to the ceiling model (*i.e.* $\lvert \boldsymbol{m}_z - \boldsymbol{p}_c \rvert_z$ where $\boldsymbol{m}_z$ is the point with the maximum $\vec{z}$ value, and $\boldsymbol{p}_c$ is a point on the ceiling)
Width	$\mathcal{W}_v$	the length along the biggest principal component, excluding $\vec{z}$

Table 7.3 Level-2 features for furniture candidates

Feature	Notation	Description
Height	$\mathcal{H}_f$	the height of the furniture candidate
Width	$\mathcal{W}_f$	the width of the furniture candidate
Nr. handles	$\mathcal{N}_h$	the number of handles present on the furniture candidate
Nr. knobs	$\mathcal{N}_k$	the number of knobs present on the furniture candidate
Min distance	$\mathcal{D}_m$	the minimum distance between the center of the planar face and the closest fixture (handle or knob)

Given the proposed multi-layered feature space, the choice of selecting a good machine learning classifier is greatly simplified, as most methods would return favorable results, given good training data. Though the goal of these experiments is to find the classes of each individual planar region separately, an interesting aspect would be however to lay the basis for a further integration between adjacent planar regions into the model. This can be done by making use of a probabilistic undirected graphical method, such as Conditional Random Fields (CRF) for example, which exploits the existing contextual information.

Figure 7.15 shows the Conditional Random Fields for the three different models. Each of the models' features is used as an input variable. The variable nodes are named after the corresponding notations in Tables 7.1, 7.2, and 7.3.

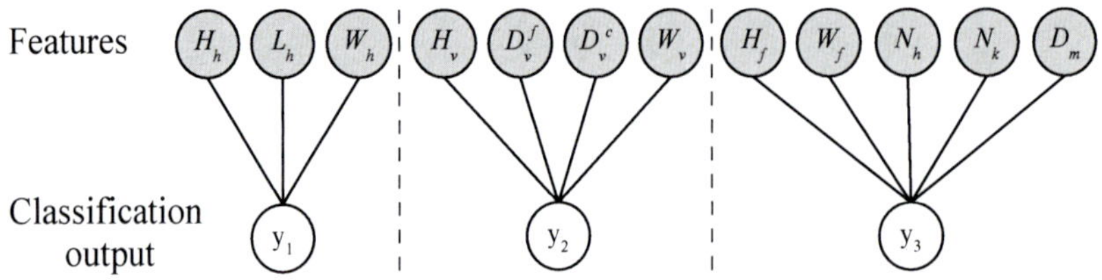

Fig. 7.15 From left to right: CRF for model 1 (Horizontal L-1 Features), 2 (Vertical L-1 Features), and 3 (L-2 Features). The classification output labels y_i are shown in Figure 7.14.

An important factor in the classification accuracy of the CRF model, is the amount and type of training data used for learning. Due to physical constraints in moving our mobile robot to a different kitchen environment, or changing the kitchen furniture to obtain multiple datasets, the amount of training data available from real world scenes is extremely limited. To circumvent this problem, we proceeded as follows: we created realistic kitchen models in the Gazebo 3D simulator, and used virtual scanning techniques, followed by synthetic data noisification to acquire additional point clouds representing kitchen environments (see Figure 7.16). After acquiring a few of these datasets, we processed them through the proposed pipeline and extracted the 2-layered features for training the CRF model. Figure 7.17 presents the planar segmentation results for such a dataset, while in Table 7.5 we show more examples of virtually scanned kitchen environments (left) and their respective fixtures on furniture candidate faces segmentation (right).

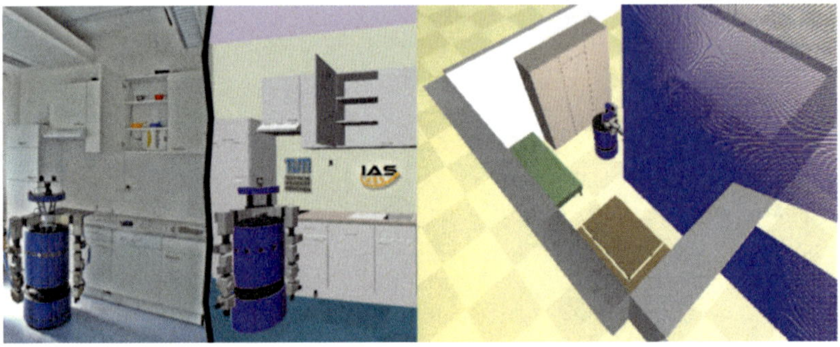

Fig. 7.16 An illustration of the simulated 3D environments and the process of acquiring training datasets, using the Gazebo 3D simulator

The classification results of the trained CRF model for the kitchen dataset presented in Figure 7.1 are shown in Table 7.4. The table shows the recall, precision and F1-measure values of all labels and the macro-averaged statistic of each model. The item accuracy is based on the overall correct classified items against the wrong classified items in the test data set.

Table 7.4 Performance results for the Conditional Random Field models

	Horizontal planes			Vertical planes			Furniture candidates		
Label	Prec.	Rec.	F1	Prec.	Rec.	F1	Prec.	Rec.	F1
1	1.00	0.50	0.67	1.00	0.91	0.95	0.94	1.00	0.97
2	1.00	1.00	1.00	0.97	1.00	0.98	0.97	0.89	0.93
3	0.96	1.00	0.98				0.50	0.75	0.60
Macro accuracy	0.99	0.83	0.88	0.99	0.95	0.97	0.80	0.88	0.83
Item accuracy		0.97			0.98			0.91	

The labels for the models given in the above table represent (in order): floor, tables, and ceiling (horizontal planes); walls, and furniture candidates (vertical planes); respectively cupboards, drawers, and kitchen appliances (furniture candidates). As it can be seen, the lowest accuracy of the classification results is represented by the kitchen appliances. The variety in the models we trained the model with is simply too large, and the proposed layer-2 features cannot capture the modeling process correctly.

Fig. 7.17 Planar segmentation results for a virtually scanned dataset using the Gazebo 3D simulator

7.3 Exporting and Using the Models

With regards to time constraints, we view the acquisition of a Semantic 3D Object Map as part of the deployment of a robot in a new environment, thus it only has to be done once before the robot starts performing its job. Once an initial model is built, changes in the map are easier to incorporate and require less time to compute.

The system stores the final object map model into a hierarchical XML-like format. This simplifies model queries and allows the map to be automatically imported

Table 7.5 Virtually scanned training datasets

Virtually scanned environment	Segmentation and model fitting
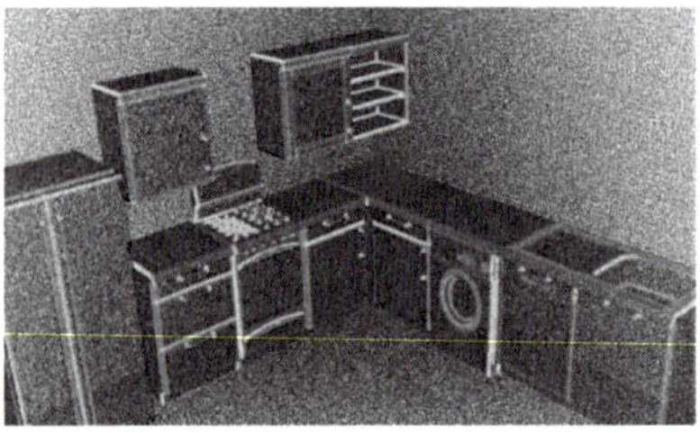	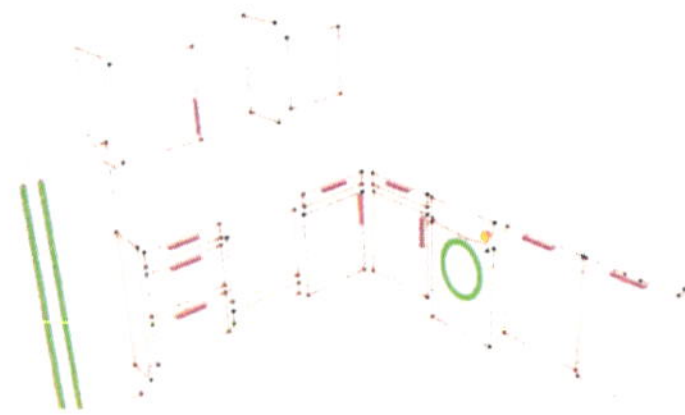
	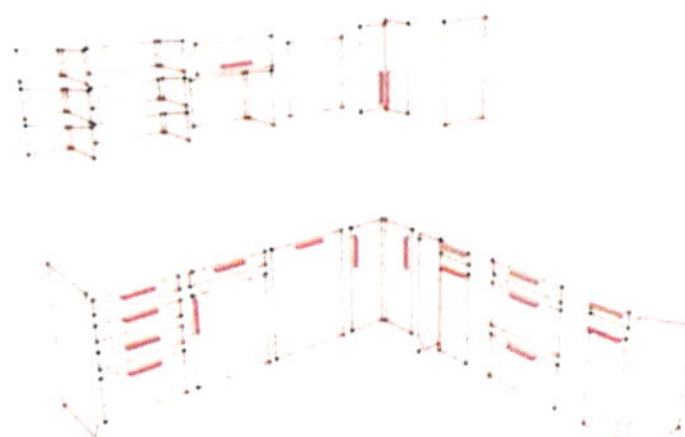
	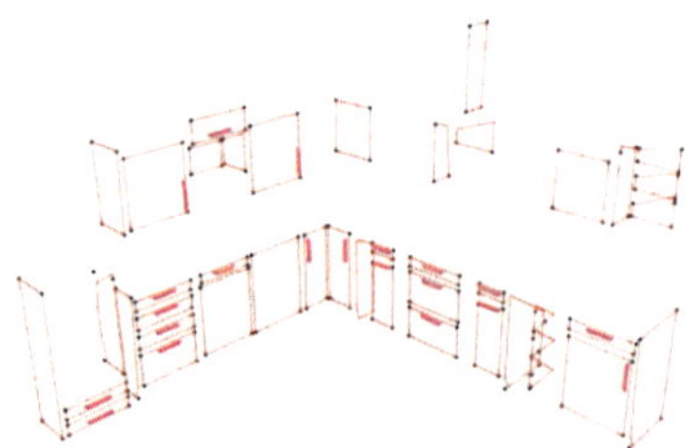

into a 3D simulation environment (Gazebo in this case), as seen in Figure 7.18. In our previous work, we have used such models of kitchens for task planning and optimization simulations [BBK^{+}07].

To support mobile manipulation and 3D collision avoidance, the proposed mapping pipeline creates a second type of map comprised of triangular meshes: the Triangulated Surface Map [MRB09]. By using the acquired object classes, the surface reconstruction methods can be applied in parallel on each object separately, leading to the creation of a *decoupled* triangle mesh map. The straightforward advantages of such a representation (see Figure 7.19 for an example) are that: a) changes in the

Fig. 7.18 Left: automatic environment reconstruction of the real world dataset from Figure 7.1 in the Gazebo 3D simulator; right: the estimation and evaluation of door hinges from geometry data

Fig. 7.19 Surface reconstruction example with mesh decoupling for all furniture candidates and objects supported by planar areas

world can be now be modeled separately on a subset of objects without loading or working with the rest; and b) it supports environment dynamics natively, as picking up an object from a table simply means moving the triangular mesh representing the object from the table into space, without the need to recreate it.

Using the sets of fixtures determined from the segmentation of point clusters lying on the vertical faces of furniture candidates, we can create representations such as the ones presented in Figure 7.20. The estimated positions of the handles can then be validated using the mobile manipulation platform. Figure 7.21 presents an example where a higher level task executive gives the robot a command to search for a particular object in a drawer. The robot navigates towards the estimated position of the handle, and attempts to open the drawer by grasping on the handle and pulling backwards. Though for the purpose of these experiments the trajectory of the robot arm was fixed a priori, the system can easily incorporate more complex motion planners to achieve a smoother path.

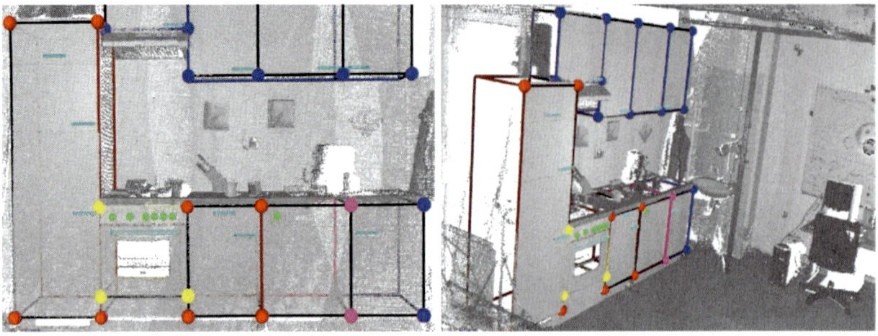

Fig. 7.20 Over imposing the resultant furniture candidates on the original point cloud data

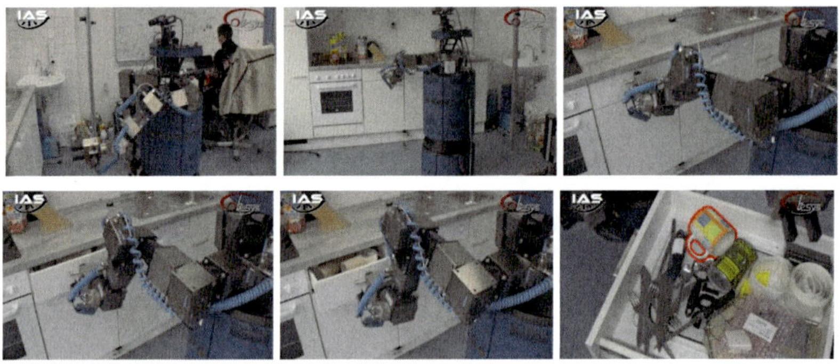

Fig. 7.21 Examples from the execution of a plan involving the opening of a drawer with the mobile manipulation platform, using the Semantic 3D Object Map

7.4 Summary

This chapter presented a comprehensive system for the acquisition of hybrid Semantic 3D Object Maps for kitchen environments. The proposed mapping system includes 2 components, namely: i) a Static Functional Map which contains those parts of the environment with fixed positions and utilitarian functions (walls, floor, kitchen appliances, cupboards, tables, etc); and ii) a Triangulated Surface Map created using decoupled triangular meshes. The former is built either by employing a set of commonsense heuristic rules on the data or by classifying a set of planar regions using simple 3D geometric features, and serves as a semantic resource for an assistant mobile personal robot, while the latter supports 3D collision detection and path planning routines for safe navigation and manipulation.

Chapter 8
Surface and Object Class Learning

> *"Objects can be classified scientifically into three major categories: those that don't work, those that break down and those that get lost."*
>
> RUSSELL BAKER

Segmenting and interpreting the surrounding environment that a personal robot operates in from sensed 3D data is an important research topic. Besides recognizing a certain location on the map for localization or map refinement purposes, obtaining accurate and informative object models is essential for precise manipulation and grasping. Although the acquired data is discrete and represents only a few samples of the underlying scanned surfaces, it quickly becomes expensive to store and work with. It is therefore imperative to find ways to address this dimensionality problem and group clusters of points sampled from surfaces with similar geometrical properties together, or in some sense try to "classify the world". Achieving the latter annotates sensed 3D points and geometric structures with higher level semantics and greatly simplifies research in the aforementioned topics (*e.g.* manipulation and grasping).

In general, at the 3D point level, there are two basic alternatives for acquiring these annotations:

1. Use a powerful, discriminative 3D point feature descriptor, and learn different classes of surface or object types, and then use the resultant models to classify newly acquired data;
2. Use geometric reasoning techniques such as point cloud segmentation and region growing, in combination with robust estimators and non-linear optimization techniques to fit geometric primitive shapes to the data. Examples include linear (*e.g.* planes, lines) and non-linear (*e.g.* cylinders, spheres) shapes but also higher order bivariate polynomials, etc.

Both approaches have their advantages and disadvantages, but in most situations machine learning techniques – and thus the first approach, will outperform techniques

R.B. Rusu: *Semantic 3D Object Maps for Everyday Robot Manipulation*, STAR 85, pp. 109–136.
DOI: 10.1007/978-3-642-35479-3_8

purely based on geometric reasoning. The reason is that while simple linear models such as planes can be successfully found, fitting more complicated geometric primitives like cones for example becomes very difficult, due to noise, occlusions, and irregular density, but also due to the higher number of shape parameters that need to be found (increased complexity). To deal with such large solution spaces, heuristic hypotheses generators, while being unable to provide guarantees regarding the global optimum, can provide candidates which drastically reduce the number of models that need to be verified.

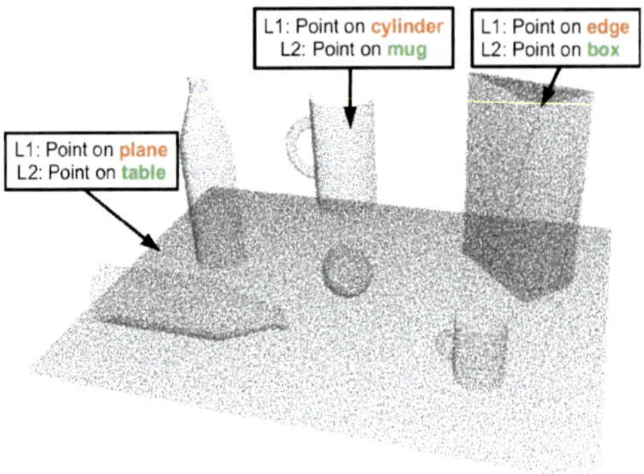

Fig. 8.1 An ideal 3D point based classification system providing two different point labels: the geometry (L1) and the object class (L2) that a point belongs to

Additionally, we define a system able to provide two different hierarchical point annotation levels as an *ideal point classification system*. Because labeling objects lying on a table just with a class label, say *mug* versus *cereal box*, is not enough for manipulation and grasping (because different mugs have different shapes and sizes and need to be approached differently by the grasp planner), we require the classification system to annotate the local geometry around each point with classes of 3D geometric primitive shapes. This ensures that besides the object class, we are able to reconstruct the original object geometry and also plan better grasping points at the same time. Figure 8.1 presents the two required classification levels for 3 points sampled on separate objects with different local geometry.

The following sections describe applications based on the first technique, that of learning surface classes using point feature representations. In particular, Section 8.1 presents work on surface labeling with Point Feature Histograms representations, followed by an optimized FPFH-based classification in Section 8.2, while in Section 8.3 we tackle the problem of object class learning using a new global descriptor, the Global Fast Point Feature Histogram (GFPFH).

8.1 Learning Local Surface Classes

This section investigates the usage of Point Feature Histogram representations for the characterization of the surface geometry on which sampled points lie, and presents an in-depth analysis on their usage for efficient point cloud classification. These classifiers can be used as accurate and reliable labeling procedures which segment point clouds into candidate regions for parameterized geometric primitive fitting. Using point labels like: *point on cylinder*, *point on torus*, etc. and the geometric relationships between them, the recognition and segmentation of objects in real world scenes can be greatly sped up and improved.

Such classification examples are given in Figures 8.2 and 8.3, where two distinct datasets acquired in an indoor kitchen environment have been processed and labeled according to four major geometric primitive surface classes. Notice how points on the two cups are labeled as cylinders (body) and edge (handle), and the surface they are lying on (table top) as planar. Making use of the resultant point classes can provide obvious advantages for applications which require good point correspondences.

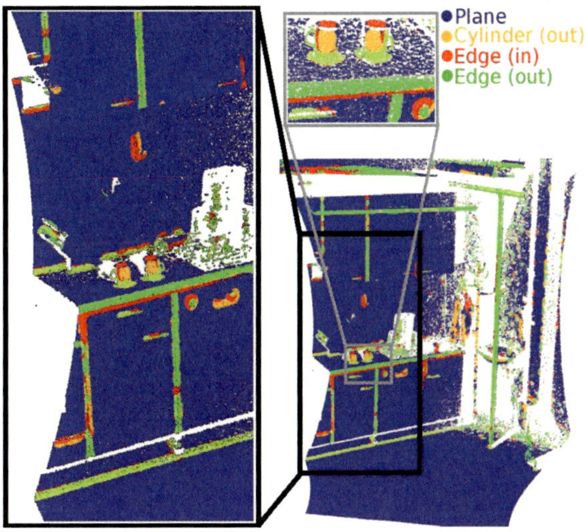

Fig. 8.2 Classification results for a dataset acquired in a kitchen environment. *In* and *out* represent concave and convex surface types.

Once the feature space is fixed, a set of classes for point labeling can be defined. One example is to define this set as representations for points lying on various 3D geometric surfaces, such as: planes, spheres, cylinders, cones, tori, edges and corners. The last two classes make sense for data acquired in indoor environments, which contains a high number of such geometric shapes instances.

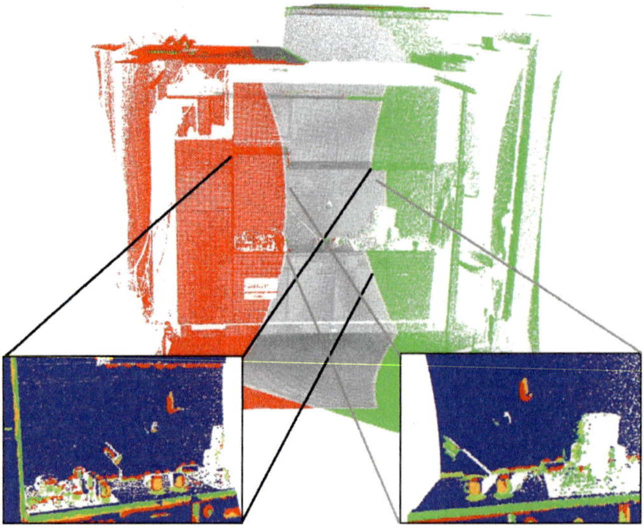

Fig. 8.3 Example of point classification for two distinct (red and green) indoor kitchen datasets registered in the same coordinate system (overlapping area shown in gray). The color legend for the classified point labels is given in Figure 8.2.

The next step is to generate training data for all possible classes. In general, using training data from real world scans has been proven to be cumbersome, because it involves a lot of manual labor into segmenting the appropriate surfaces for each class. Moreover, for non-linear shapes (cones, tori, etc) this manual segmentation is prone to errors. A viable solution is to generate synthetic data that represents the desired primitive surfaces, and apply the same noise levels found in the real-world datasets to it. The classification results can be then tested using marked (ground truth) point clouds but also visually on real-world datasets where this information is missing.

In the context of the chosen application, the selection of r_2 for the computation of PFH representations (see Section 4.4) should be adjusted to the size of the shapes that need to be detected and to the level of noise that is expected from the sensor. The value of r_1 should be big enough to balance the effect of noise, but small enough to preserve the local nature of the estimated surface normal. Similarly, the value of r_2 should be big enough to capture enough information about the shape, but small enough to avoid the overlapping effects of multiple surface types in the same neighborhood.

In indoor scenes containing objects of every day use (*e.g.* cups, bottles, cereal boxes, etc), the shape primitives used for training can be generated to have appropriate dimensions with respect to these objects, with examples including r_1=1.5cm and r_2=2.5cm for dense laser scans. These values are required for an accurate surface representation for a cup's handle for example, which constitutes the smallest shape (torus) that is to be detected given the noise level found in the input dataset [RMBB08b].

8.1.1 Generating Training Data

Since the chosen PFH hyperspace requires consistently oriented normal information and bases its computational model on that, there are two types of situations that need to be accounted for:

- when the estimated surface normals of a shape are oriented towards the concave (inner) part of the surface (Figure 8.4 left);
- when the estimated surface normals are oriented towards the convex (outer) part of the surface (Figure 8.4 right).

The above does not imply that the PFH feature space is pose variant, but that it identifies the way an object is oriented to some degree with respect to the viewpoint (*e.g.* a corner of the room will be marked differently than a corner of a piece of furniture in the room – inner and outer corner respectively). Figure 8.4 presents the normal information for a cylinder in both the above mentioned cases.

Fig. 8.4 Estimated and oriented surface normals for a cylinder pointing towards the concave (left) and convex (right) part of the surface

Therefore, the number of surface classes has to be multiplied by two, with the exception of the plane. Figure 8.5 presents Point Feature Histograms for points lying on the selected geometric shapes, when the surface normals are oriented towards the concave (left) and convex (right) parts of the shape.

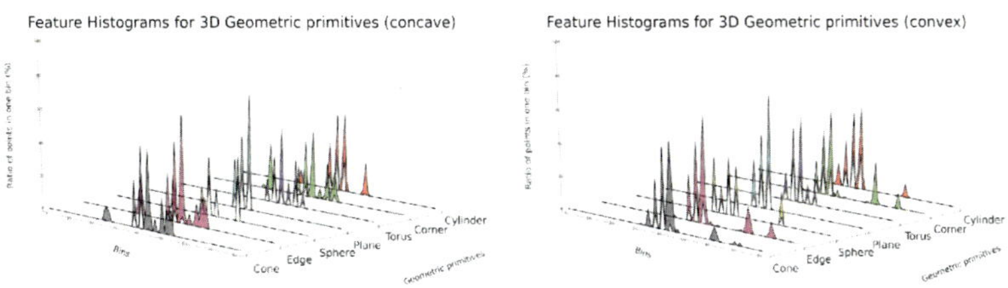

Fig. 8.5 Point Feature Histograms for points lying on several 3D geometric primitives: normals oriented towards the concave (left) and the convex (right) parts of the shape

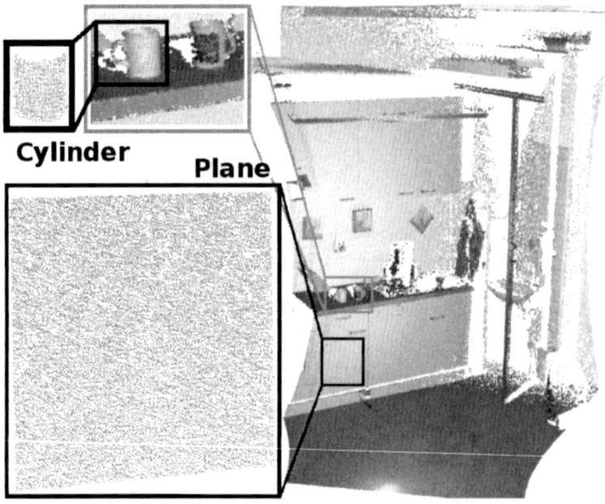

Fig. 8.6 Planar and cylindrical surface patches used for noise analysis

The presented histograms were computed for shape primitives that were generated synthetically and are noiseless. The results show that the different geometrical properties of each surface produce unique signatures in the Point Feature Histograms space. It is important to mention that the point density on the surface does not influence the features significantly, however the points on the edges of the generated shape primitives have small differences compared to those in the middle of the surface, producing the variations that can be seen in Figure 8.5.

An important requirement of the selected feature space is that it has to be able to cope with noisy datasets, acquired using real hardware sensors. To see whether the point histograms will be discriminating enough, we will analyze the level of noise in a scanned point cloud, add the same level of noise to our synthetic datasets, and compare the resultant histograms. The particular datasets presented here have been acquired using a SICK LMS laser sensor in an indoor kitchen environment (see Section 2.1).

For determining accurate noise models, two portions of a scan (see Figure 8.6) have been manually segmented to represent both linear shapes (*e.g.* planes) and non-linear (*e.g.* cylinders). To compute the noise parameters for the underlying shape model, the following steps have been taken:

1. Using a Sample-Consensus based method (MSAC [TZ00]), a search for the best shape support (plane or cylinder) is performed;
2. A linear least-squares fit for the plane and a non-linear Levenberg-Marquardt optimization for the cylinder is used to refine the obtained model, and compute the shape parameters;
3. The distances from all the points in the dataset to the estimated shape model are estimated and used to build a Gaussian distribution of distances with a mean μ and a standard deviation σ.

By analyzing the statistics of the distribution and comparing the results with the noiseless case, we can deduct, with some precision, the actual noise levels present in the input scanned data. The left part of Figure 8.7 presents the distance distributions of the cylinder's points from the computed underlying shape model for the cylinder shape. The resulted distance distribution for the planar patch has practically the same mean and standard deviation values as the ones for the cylinder. In the figure, μ_{raw} and σ_{raw} represent the noise mean and standard deviation for the raw (scanned) dataset. By adding the same level of noise to the synthetically generated datasets and repeating the above computational steps, a new distance distribution with similar mean and standard deviation values, μ_{syn} and σ_{syn} (see the right part of Figure 8.7) can be obtained. The fitted Gaussian distribution is shown in black, the $\pm\sigma$ standard deviation in red, and the mean μ in blue. The "noisification" procedure can be applied on all three dimensions or alternatively only on the $\vec{z}$ (*i.e.* depth) axis.

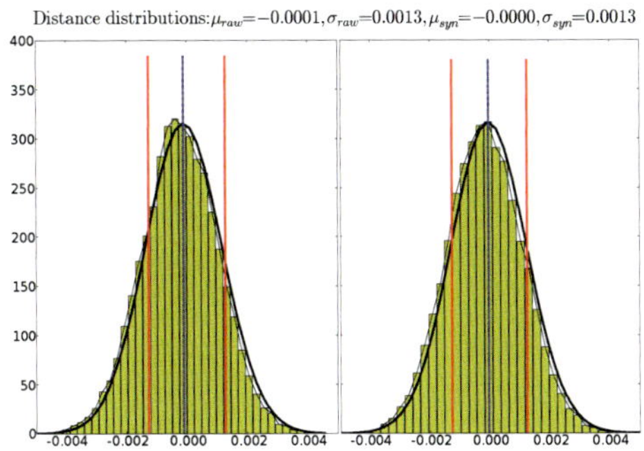

Fig. 8.7 Point to surface distance distributions for a point cloud representing a cylinder: raw scan (left), synthetic dataset with added noise (right)

Figures 8.8 and 8.9 show the Point Feature Histograms for points lying on planar and cylindrical surfaces: synthetically noiseless generated data on top, the noisy raw dataset in the middle, and the synthetic dataset with added noise on the bottom.

This analysis allows for fine refinements of the estimated PFH representations such that they resemble the raw scanned data as much as possible. As shown, the resultant noisified synthetic data captures the underlying geometry in good detail, thus reducing the need for a manual segmentation for the generation of training data in the model learning process.

8.1.2 Most Discriminative Feature Selection

After obtaining enough datasets representing the geometric primitives of choice, and after their appropriate features have been computed and extracted, a selection of the most *relevant* features has to be performed. The selection is motivated by the fact that for a given shape, there could be tens of thousands to millions of such training instances, which, if used directly, would render the learning problem costlier than

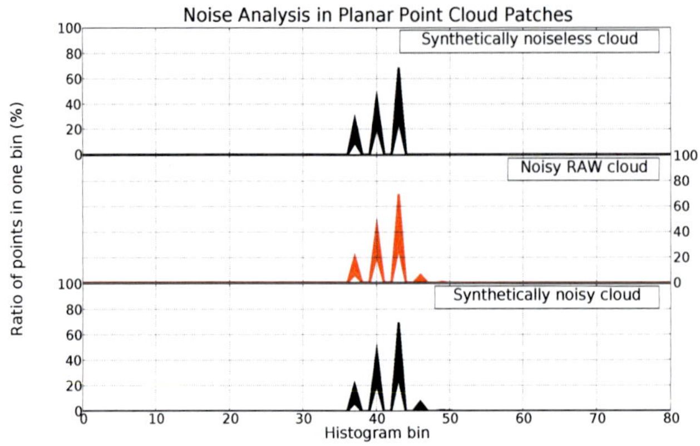

Fig. 8.8 Point Feature Histograms for points on a plane for synthetic noiseless data (top), raw point cloud data (middle) and synthetic noisy data (bottom)

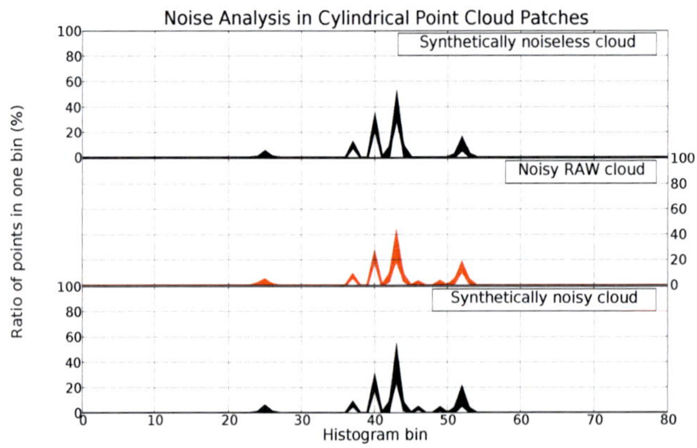

Fig. 8.9 Point Feature Histograms for points on a cylinder for synthetic noiseless data (top), raw point cloud data (middle) and synthetic noisy data (bottom)

needed. Therefore, a subset of features which reduces the number of training examples but still preserves the same information with respect to the shape signature, has to be identified and extracted.

This can be viewed as a clustering problem, in the sense that the goal is to find the most relevant k clusters in the hyper-dimensional PFH space, which taken together resemble the overall shape representation without losing or modifying its discriminative properties. A solution to this problem is to treat the clustering in an unsupervised learning formulation, and thus apply an algorithm such as K-Means, where k (the number of clusters) could be determined empirically.

However, since the proposed feature space is hyper-dimensional, it is very likely that the iterative clustering method will be trapped in local minima, and the optimal cluster centers will not be found. A solution is to reduce the dimensionality of the space and perform the clustering there. However, even in a lower dimensional space, the risk of converging to a false solution is directly related to the initial starting position of the clusters.

The solution proposed here consists in computing the mean μ-histogram of the geometric shape, then creating a distance distribution from all the point feature histograms to it, divide it into equally distributed parts, and select a representative from each part as a starting solution for the k cluster centroids. In detail, the procedure for determining the most discriminating features is represented by the following computational steps:

1. For each shape candidate, compute the mean μ-histogram of the shape;
2. For each point in the shape, compute a distance metric D_m between the point's feature histogram and the mean μ-histogram of the shape;
3. Arrange all the distance values in a distribution, divide it into k intervals, and uniformly sample an index value from each interval;
4. Use the sampled distances as initializers for the k center clusters, and search for an optimal solution using a K-Means formulation.

The proposed method gives similar results to the one proposed in [Vas03]. For determining the optimal D_m, we have performed an in-depth analysis using multiple distance metrics and norms (such as L1-Manhattan, L2-Euclidean, Bhattacharyya, Chi-Square, Jeffries-Matusita, Kullback-Leibler), similar to the ones presented in Section 4.5. The empirical evaluation indicated good results using the L1-Manhattan metric.

Figure 8.10 presents the results of the selection process for a conic shape with 10815 feature histograms. For the purpose of these experiments, the number of k clusters was set to 40. In practice, we require the same number of training examples for each candidate shape. To achieve this, we first compute the feature histograms for all shape primitives, and then we select:

$$k = \min(\alpha, n^{s}{}_{i}),\ i \leq N \tag{8.1}$$

where n^s is the number of feature histograms for shape s, N is the total number of primitive shape classes, and α is a maximum tolerated number of clusters set by the user.

Selecting the most discriminating features using the proposed K-Means clustering approach has the advantage that the learning time will be drastically reduced while the discriminating feature information will be preserved for training the machine learning classifier.

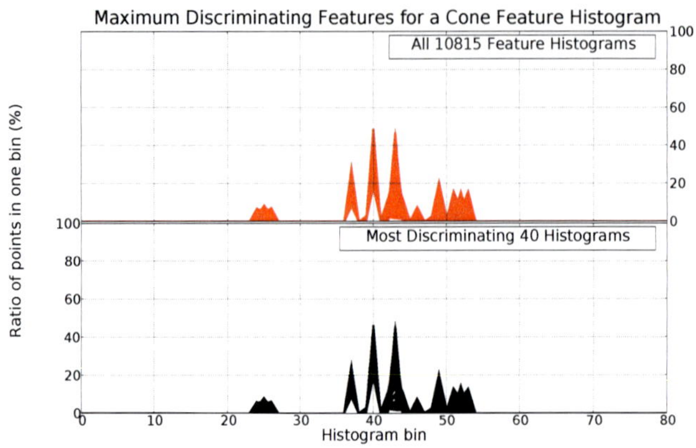

Fig. 8.10 Most discriminating 40 feature histograms for points lying on a concave conic surface. All 10815 feature histograms are shown at the top of the figure, and the resultant 40 feature histograms at the bottom.

8.1.3 Supervised Class Learning Using Support Vector Machines

The goal of the application presented herein is to learn models for 3D geometric primitives (see Figure 8.5) and then apply these models to classify scanned data coming from laser sensors. The separation of the PFH hyperspace into distinct classes can be reformulated as a Support Vector Machine classification problem. Given a set of training examples, the SVM learner attempts to find optimally separating hyperplanes which group sets of training examples belonging to the same class and separate those from different classes. The support vectors are usually found in a higher dimensional feature space where the data is mapped by making use of a kernel function. Choosing or designing a new kernel function is an important decision, as selecting the wrong kernel can lead to inadequate results.

By applying the most discriminating feature selection algorithm as presented in Section 8.1.2, the training datasets become less voluminous. Trying out different kernels and optimally determining their best coefficients in a large parameter space therefore becomes possible by training different models in parallel and selecting the best one. Besides the classical linear, sigmoid, and polynomial kernels, and motivated by the results in [CHV99], we have selected the family of Radial Basis Functions kernels:

$$K_{RBF}(x,y) = e^{-\gamma \cdot dist(x,y)} \tag{8.2}$$

By changing the distance metric, the kernel can take the form of a Laplacian, Gaussian or sublinear kernel (see Appendix C.1). Both the Laplacian and the sublinear kernels have shown great results for histogram matching [JNY07, CHV99], outperforming the Gaussian kernel. Supervised learning techniques such as SVM have the advantage that while the model learning is slow (solved here through the use of the most discriminating feature selection), the classification of new unseen instances is performed very fast.

The problem of classifying new point features can also be formulated in a different manner, that is without directly specifying the target output classes. This falls into the context of semi-supervised or unsupervised learning techniques, and its main advantage is that the learning process does not require sets of labeled data to be used for training the model. A rather effective and simple histogram classification method can be obtained by looking at different distance spaces and performing a K-Nearest Neighbor (KNN) search for the closest match, where $k = 1$. That is, for every shape training set, compute the mean μ-histogram, and assign the shape class to a point $\boldsymbol{p}$, if the point's histogram $\boldsymbol{p}_i^f$ has the closest distance to it than to any other mean shape histogram. In the context of the application presented in this chapter, we evaluated the following distance metrics: Manhattan (L1), Euclidean (L2), Bhattacharyya, and Chi-Square (χ^2) (see Appendix A.2 for their equations).

A similar method for classification is K-Means clustering, where instead of assigning the shape class directly as KNN does, an iterative search for the optimal k clusters is performed. Given a good metric space histograms belonging to the same class tend to be grouped together in the same cluster.

Table 8.1 presents the results obtained using the above classification methods and different distance metrics for a synthetically generated scene containing various geometric shapes (see Table 8.2). The best classification support for both noiseless and noisy data with Support Vector Machines was obtained using the RBF sublinear kernel, confirming the findings in [CHV99]. The RBF Laplacian kernel gave similar results, albeit the RBF Gaussian kernel provided better results for noisy data. An interesting result was obtained using the linear kernel, which had the second best results for noiseless data. This means that when a lower computational cost is desired with small acceptable penalty losses in performance, a linear SVM model could be used instead. Note that the numbers presented in the table are directly related to the synthetic dataset used – if ground truth would be available for real datasets, the outcome of the classification might result in bigger differences between the kernels.

Table 8.1 Classification results (in %)

Method used	**Noiseless**	**Noisy**	**Method used**	**Noiseless**	**Noisy**
SVM Linear	95.17	87.66	KNN L1	78.08	78.03
SVM Polynomial	94.39	88.88	KNN L2	67.22	71.94
SVM Sigmoid	86.15	83.44	KNN Bhattacharyya	**87.11**	**83.53**
SVM RBF Gaussian	94.55	88.83	KNN Chi-Square	83.64	82.84
SVM RBF Laplacian	95.18	88.14	K-Means L1	**73.63**	68.58
SVM RBF Sublinear	**95.26**	**89.55**	K-Means L2	61.30	68.58
K-Means Bhattacharyya	**73.63**	**70.74**	K-Means Chi-Square	**73.63**	68.58

Table 8.2 presents the classification results using the best parameters with each of the three methods (SVM, KNN, and KMeans) for the synthetic noiseless scene (left) and for the scene with added noise (right). In the legend presented at the bottom of the table, two types of tori were included for detecting cups handles: an ideal torus (*rounded*) type, and another type more *flat*. Instantiations of the two different types of handles can be seen in Figure 8.11.

The KNN based classification gave very promising results, especially when using the Bhattacharyya distance metric. This is due to the fact that the discriminating power of the PFH representations is high enough so that in certain distance spaces they can be easily separated. The presented results make the KNN method very attractive, especially for situations where a model does not have to be learned a priori. The K-Means clustering algorithm had the worst results of the three methods. Even though the k cluster centroids were initialized as close as possible to the true solution, the final solution diverged from the true class, due to the iterative nature of the algorithm and the metric space used.

Another aspect that needs to be verified is how the PFH classification copes with partial scans taken from different positions and angles. Figure 8.11 presents the classification results for 4 types of cups scanned from 3 different poses. In the first row, the acquired scans contain the cup handles as seen from a 45° angle and slightly from above; in the second row, the handles are precisely in the middle of the scan and are perpendicular to the viewpoint, while the position is above the cup almost looking down at the inside bottom of the cup; and finally, in the third row, the handles are exactly at a 90° angle with the viewing direction, allowing them and the cup to be seen from the side. The results obtained indicate the following findings:

- all scans could identify the concave and convex cylindrical components of the cups within a reasonable error margin;
- the handles for the first 3 cups were identified as *flat* tori in situations where more than half of the handle was visible, and as *rounded* tori when seen from the side (*e.g.* the bottom row of the figure);
- the third cup had an unusual shape (*i.e.* not perfectly cylindrical) and has been classified as a composition between a normal concave cylinder (bottom), a convex cylinder (middle) and a convex torus (top), showing that the classification of more complex objects is possible, but has to be interpreted differently.

To verify the robustness of the PFH representations on a more complex dataset where objects are grouped closer and occluding each another, we analyzed the scene presented in Figure 8.12 and obtained 90.26% classification results using a sublinear SVM kernel. Notice that our ground truth labels were expecting edges at the intersection of shape candidates, such as cylinder with plane, or sphere with plane, as well as torus with cylinder. Due to the r_2 value used, the torus representing the handle of the smaller cup was not classified as a torus, but as an edge. This leads to the conclusion that the overall classification accuracy could be improved if different levels of details are used (*i.e.* different radii r_1 and r_2).

Table 8.2 Classification results for the synthetic scene. From left to right: noiseless and noisy synthetic scenes. From top to bottom: SVM sublinear kernel, KNN with Bhattacharyya distance, and K-Means with Bhattacharyya distance.

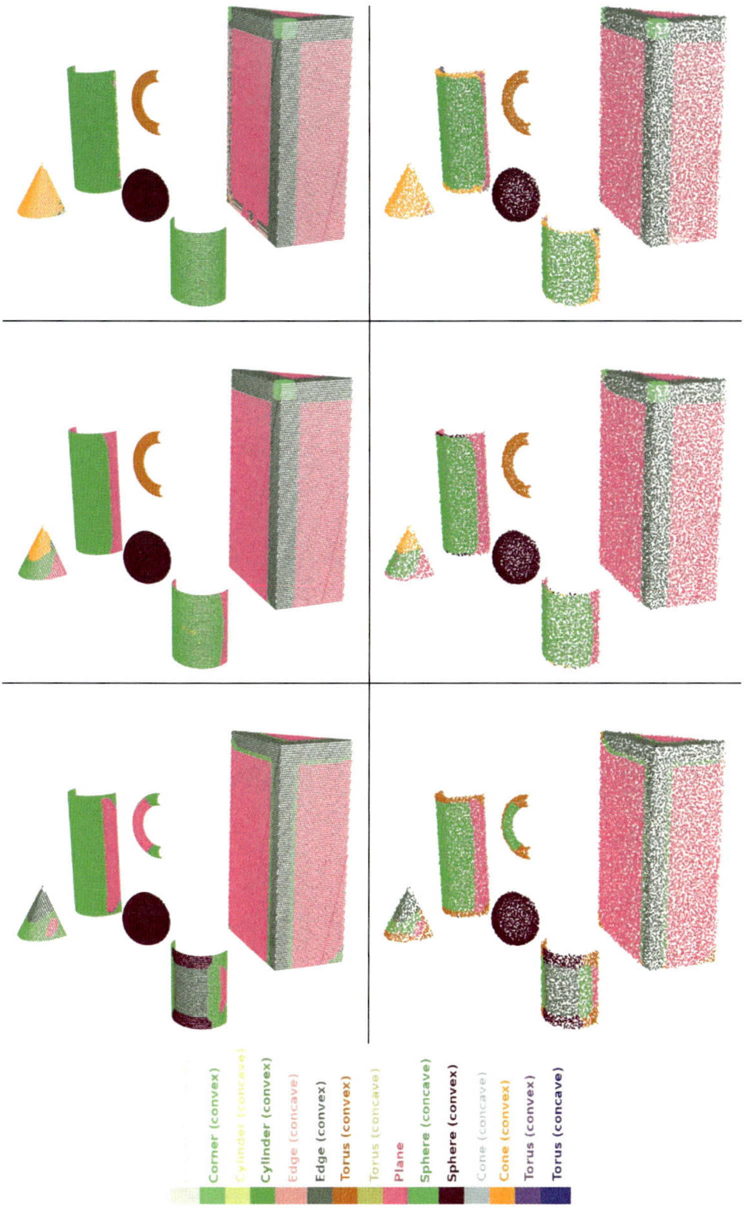

Fig. 8.11 Classification results for 4 types of cups scanned from different positions and angles

Fig. 8.12 Complex noisy synthetic scene with 90.26% classification results using a sublinear SVM kernel

Fig. 8.13 Point Feature Histogram classification results for a table scene with kitchen objects shown as an intensity image (left) and the results after classification (right)

The results obtained by applying the same SVM model to a real-world scene are shown in Figure 8.13. The left part of the figure presents the point cloud in intensity scale, while the right shows the classification results. Similar other datasets have already been presented in Figures 8.2 and 8.3.

8.2 Fast Geometric Point Labeling

The results presented in Section 8.1 have shown a good separation of points based on the underlying properties of the geometric surface they are sampled on, captured by their respective Point Feature Histogram representations. Their major disadvantage is connected to the estimation complexity of the PFH features, which for a point cloud dataset $\mathcal{P}$ with n points is $O(nk^2)$, with k being the number of neighbors of each $\boldsymbol{p}_i \in \mathcal{P}$ that we need to estimate the representations for. This makes the PFH based classification system unsuitable for application with tighter computational constraints.

In the following, we devise a similar classification scheme based on the less demanding FPFH representations (see Section 4.5) and probabilistic graphical methods. The switch from Support Vector Machines to probabilistic graphical methods such as Conditional Random Fields is justified by the incapability of SVM representations to enforce the same class labels to the surrounding neighbors of a point. As it can be seen in most of the figures from Section 8.1, there are tiny groups of points in several areas of an object or surface which appear as being labeled with a totally different class than the rest of the points in that area. In contrast, discriminative undirected graphical models can make use of contextual information and enforce the labeling of adjacent points to the same class label, as presented in Appendix C.2.

As an application scenario, we want to demonstrate the parameterization and usage of a FPFH-based system for the purpose of point classification for objects lying on tables in indoor environments using Conditional Random Fields [RHBB09]. Figure 8.14 presents a classification snapshot for a scanned dataset representing a table with objects in a kitchen environment. The system outputs point labels for all objects found in the scene.

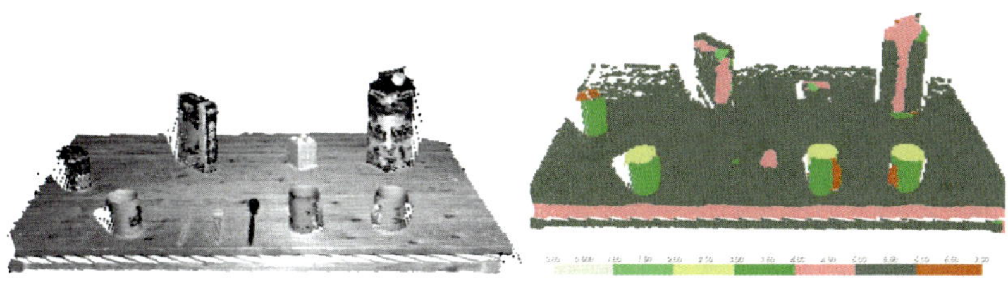

Fig. 8.14 Left: raw point cloud dataset acquired using the platform in Figure 8.16 containing approximatively 22000 points; right: classification results for points representing the table and the objects on it. The color legend is: dark green for planar, mid green for convex cylinders, yellow for concave cylinders, light green for corners, pink for convex edges and brown for convex tori.

The given problem of labeling neighboring points motivates the exploitation of the graphical structure of Conditional Random Fields. In our work, each conditioned node represents a 3D point and each observation node a calculated feature. Figure 8.15 shows a simplified version of the model, with the output nodes connected to their observation nodes and the observation nodes of their surrounding neighbors.

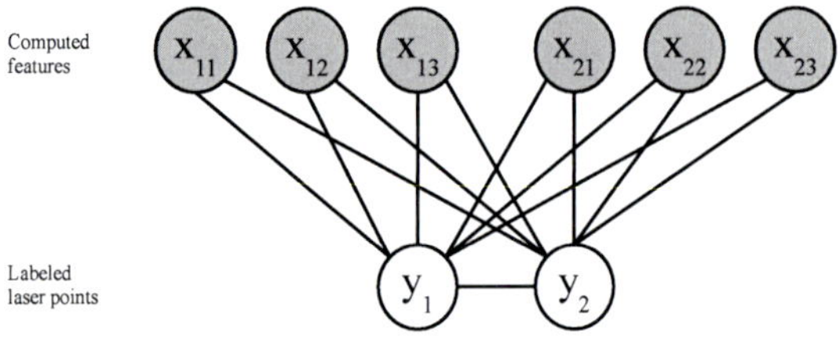

Fig. 8.15 A simplified Conditional Random Field model for FPFH based classification

The validation of the FPFH based framework was done using the mobile manipulation platform presented in Figure 8.16. By mounting a SICK LMS 400 laser scanner on the robot arm, and sweeping the scenes with a frequency of 1Hz, we acquired a large number of datasets representing table setting scenes. Each point cloud dataset comprises around 70000 – 80000 points.

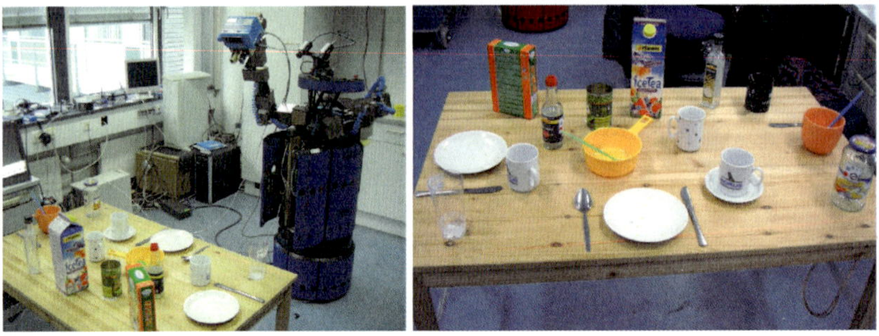

Fig. 8.16 Left: the mobile manipulation platform used in our experiments; right: an example of one of the scanned table setting scenes

Since we are only interested in classifying the points belonging to the objects on the table, the raw point clouds are prefiltered and the table surface is segmented together with the individual point clusters supported by it. To do this, we assume that the objects of interest we are interested in are lying on the largest horizontal

Fig. 8.17 Left: raw point cloud dataset; right: the segmentation of the table (green) and the objects supported by it (random colors) from the rest of the dataset (black)

planar components in front of the robot. The estimation of these horizontal planes is done in a MSAC [TZ00] framework, as presented in Chapter 6. Once the model is obtained, the algorithm estimates the boundaries of the table as a 3D polygon and segments all object clusters supported by it (see Figure 8.17). Together with the table inliers, this constitutes the input to the FPFH feature estimation and surface classification experiments, and accounts for approximatively 22000-23000 points ($\approx 30\%$ of the original point cloud data).

To learn a good model, object clusters from three different data sets have been segmented and manually labeled. This is in contrast to the previous formulation from Section 8.1, where the classifiers were trained with synthetically noisified data. The training set was built by choosing primitive geometric surfaces out of the data sets, such as: corners, cylinders, edges, planes, or tori, both concave and convex. The initial estimate is that these primitives account for over 95% data in table setting scenes similar to the ones used for the purpose of these experiments.

To compare against the results in Section 8.1, both the PFH and FPFH representations have been estimated for each point in the dataset, and then used to train two separate Support Vector Machine and Conditional Random Fields models.

Since ground truth was unavailable during the data acquisition stage, we manually labeled the scenes to test the individual accuracy of the separate machine learning models obtained. The results with respect to computational performance and classification accuracy are presented in Table 8.3.

As shown, the FPFH algorithm easily outperforms the PFH algorithm in computation performance. The estimation of the FPFH representations is over 20 times faster for the particular datasets used in these experiments. An important observation is that the CRF model based on the decorrelated FPFH representations provided better results than the PFH based model. This can be explained by the lesser influence of neighboring points for the feature representation of the query point in the FPFH formulation.

The training error curves for the CRF models for both PFH and FPFH are shown in Figure 8.18. Since the error norm becomes very small after approximatively 50 iterations, the model training can be stopped there. Table 8.4 presents visual examples of some of the classified table scenes with both the SVM and CRF models.

Table 8.3 Feature estimation, model learning, and testing results for the dataset in Figure 8.14. The method indices represent the type of features used: PFH (^{1}) and FPFH (^{2}). All computations were performed on a Core2Duo @ 2 GHz with 4 GB RAM.

Method	Feature estimation (pts/s)	Model training (s)	Model testing (pts/s)	Accuracy
CRF^1	331.2	0.32	18926.17	89.58%
SVM^1	331.2	1.88	1807.11	90.13%
CRF^2	8173.4	0.091	76996.59	97.36%
SVM^2	8173.4	1.98	4704.90	89.67%

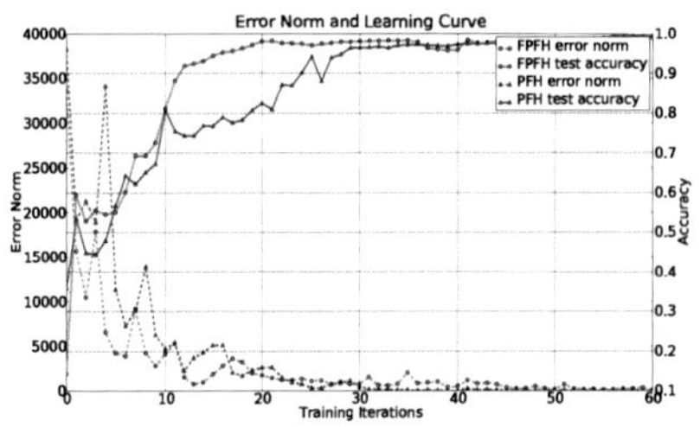

Fig. 8.18 Test accuracy and training error curves for the two Conditional Random Field models for PFH and FPFH representations

The number of FPFH representations per second that the system can estimate as given in Table 8.3, has been obtained using a single-threaded implementation. For applications with timing constraints, the algorithm can be extended to spawn multiple threads and estimate several FPFH features at once. Additionally, if the scanning data is acquired in a sweeping manner, an algorithm that can estimate the FPFH representation online while the scan is being performed can be implemented as proposed in our previous work [RBB09].

8.3 Global Fast Point Feature Histograms for Object Classification

In contrast to the two application scenarios mentioned in Sections 8.1 and 8.2, this section deals with a more complex problem in the context of learning surface and object models for mobile manipulation and grasping applications from noisy stereo data. The approach presented here makes use of dense 3D point cloud data acquired using stereo vision cameras by projecting textured light onto the scene (active stereo). To create models suitable for grasping, the supporting planes are first extracted in a manner similar to the methods presented in Section 6.4, followed by the segmentation of object clusters supported by them. All point clusters are first labeled with different surface geometric primitives using a FPFH representation at a local level. Then, a novel global descriptor (Global Fast Point Feature Histogram) is used to categorize the individual clusters into object classes using the previously estimated labels at each point. To create models useful for grasping applications, all resultant decoupled primitive point clusters are reconstructed as smooth triangular mesh surfaces.

Table 8.4 A subset of 4 different datasets used to test the learned models. The images (left to right) represent: grayscale intensity values, classification results obtained with the SVM model, and classification results obtained with the CRF model. The color legend is similar to the one given in Figure 8.14.

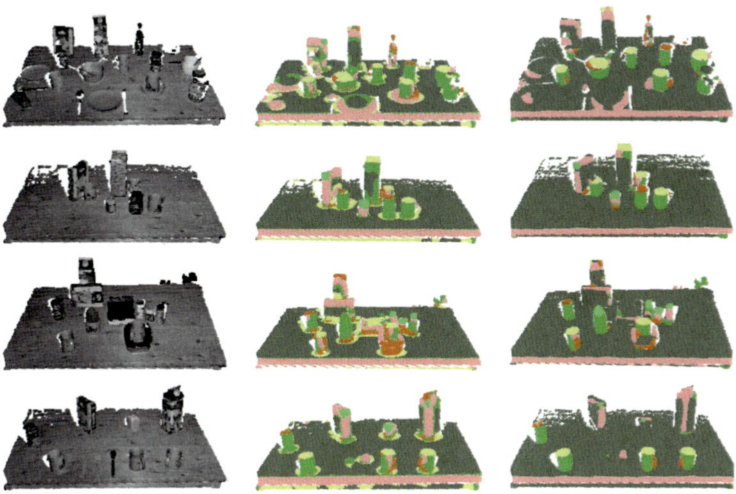

The focus of the application presented in this section is on modeling the immediate area that a robot's arms can reach, which is called the robot's workspace. Because the robot is mobile, it is impossible to instrument the environment with sensing devices such as active vision systems for example, but these could be easily installed on the robot. As previously described in Section 2.1, active stereo sensors can produce dense 3D depth maps by projecting random texture patterns onto the scene. In comparison with panning 2D laser based systems, the data can be acquired at higher frame rates, usually in the order of 30-60Hz, but at the cost of more noise.

A proposed system architecture able to learn point and object classes and reconstruct the scanned surfaces as triangular meshes from noisy active stereo data is presented in Figure 8.19.

For every point cloud $\mathcal{P}$ acquired from the stereo system, the processing pipeline first estimates the underlying surface normal estimates $\vec{\boldsymbol{n}}_i$ at every point $\boldsymbol{p}_i \in \mathcal{P}$, by approximating them with the normals of least-squares planes fit to local k-nearest neighbors patches centered around each $\boldsymbol{p}_i$. By transforming the input data into the robot coordinate system ($\vec{z}$ pointing upwards), and using the previously estimated surface normals, a parallelized segmentation scheme for all horizontal planar surfaces $\mathsf{t}_i \in \mathcal{T}$ sampled in $\mathcal{P}$ is devised using robust MSAC (M-Estimator Sample Consensus) estimators [TZ00]. Then, the closest t_i candidates to the robot are selected and all Euclidean point clusters supported by them (*i.e.* sitting on it) are extracted from the data. Since these clusters are independent with respect to each other, the next step is to estimate local Fast Point Feature Histograms at every point in a cluster, for each cluster in parallel. Using a previously learned Conditional Random

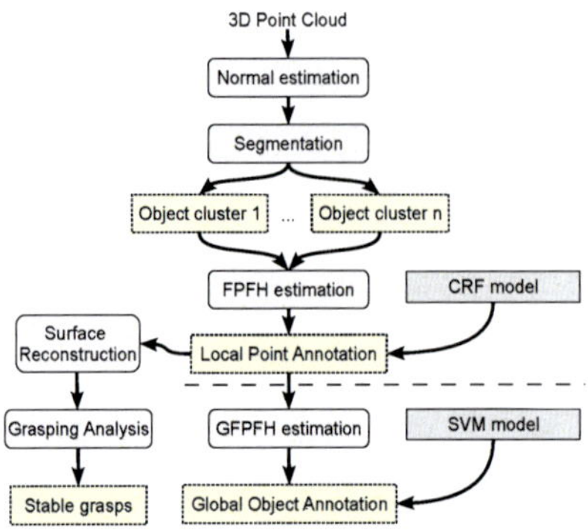

Fig. 8.19 The architecture of the proposed mapping system, with the major processing steps highlighted with light blue color. The two classification layers produce local point annotations (CRF - layer-1), and global object annotations (SVM - layer-2) respectively. The outputs of each step are represented with yellow dotted boxes.

Fields model, the system outputs local point classes. These local point annotations are then used for computing a global representation for each cluster using GFPFH descriptors, which in turn are used to train a Support Vector Machine model that classifies clusters into object classes. Additionally, using the same local point labels, a surface reconstruction step is applied to generate smooth triangular surface meshes useful for grasping applications. The grasping analysis is performed offline by estimating the force closure and other grasp stability metrics between the reconstructed surfaces and the robot grippers using approaches such as [MA04, DK08]. An example of the intermediate outputs of the proposed architecture is shown in Figure 8.20.

Fig. 8.20 From left to right: raw point cloud dataset, planar and cluster segmentation, labeling of the point cluster using local FPFH representations, and finally the reconstruction of the scene surface using smooth triangular meshes

In principle, the local point classification using Conditional Random Fields follows the structure presented in Section 8.2. The CRF model is used for predicting the point labels. The feature nodes x_i are connected to the output variable y representing the geometric classes. Since all y nodes are connected amongst each other, the same output label is enforced locally, and neighboring points will be labelled with the same geometric class. The point features are calculated using a slightly modified FPFH variant, by selecting the weight proposed in Equation 4.22 as:

$$\omega_j^k = \sqrt{\exp \|\boldsymbol{p} - \boldsymbol{p}_j^k\|} \tag{8.3}$$

The new weight is not penalizing the neighboring points on the surface too much and leads to a larger influence of the neighboring points $\boldsymbol{p}^k$ SPFH features on the resultant FPFH for the query point $\boldsymbol{p}$.

To validate the proposed framework, we have performed several experiments of point annotation using the CRF classification approach. In particular, over 500 datasets of table setting scenes have been gathered, containing objects such as the ones presented in Figure 8.21. Each dataset was first segmented using the methods described above. The remaining point clusters had on average approximatively 600 points in a radius of 3 cm.

Fig. 8.21 The complete collection of IKEA objects used for the active stereo experiments presented in this section

Due to the nature of the objects selected for these experiments, four major surface primitive classes have been defined: planar, cylindrical concave, cylindrical convex, and stem or handle (represented by thin and slightly planar patches). Though a few glasses in the dataset include conical parts, the levels of noise in the point cloud datasets made distinguishing conical from cylindrical too unreliable.

For the sake of completeness, for each point $\boldsymbol{p}_i \in \mathcal{P}$, three different feature representations have been estimated, namely: the PFH and FPFH representations as described in Sections 4.4 and 4.5, and the modified weight FPFH variant presented in this section (see Equation 8.3). The overall classification results are presented in Table 8.5, together with the total computation time required for feature estimation, model learning and testing, on a standard Intel Centrino 1.2Ghz notebook, while the

Table 8.5 Feature estimation, model learning, and overall testing results for the Conditional Random Fields model. FPFH^m represents the modified weight FPFH variant.

Method	Feature estimation (pts/s)	Model training (s)	Model testing (pts/s)	Accuracy
PFH	13.33	10.43	3936	57.51 %
FPFH	1128.22	4.84	1699	90.49 %
FPFH^m	1203.39	0.65	10087	98.27 %

Fig. 8.22 Examples of FPFH based classification for active stereo data representing scenes with multiple objects. The color legend indicates the following labels: cylindrical convex (green), cylindrical concave (red), stem and handle (blue).

training error curves are given in Figure 8.24. Figure 8.22 presents several classification examples using the trained CRF model.

Once the point labels are obtained, we proceed at reconstructing the original surfaces using a decoupled surface mesh triangulation [MRB09], followed by a simple mesh relaxation [TZG96, Lin00] to obtain smoother surfaces and normals. Figure 8.23 presents several results of the surface reconstruction step for different objects located on planar surfaces. The resolution of the triangular mesh is directly dictated by the requirements of the grasping analysis step.

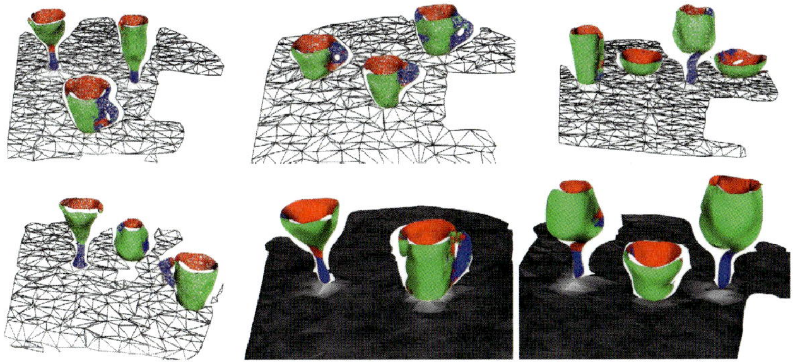

Fig. 8.23 Examples of the resultant triangular surface meshes using our reconstruction pipeline. Note that each object is individually reconstructed, thus allowing the surface reconstruction algorithms to be run in parallel.

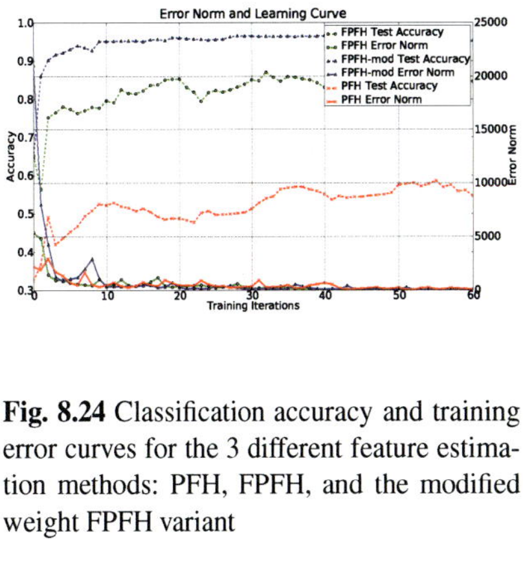

Fig. 8.24 Classification accuracy and training error curves for the 3 different feature estimation methods: PFH, FPFH, and the modified weight FPFH variant

Figure 8.25 presents the geometric processing steps for a more complex scene, with objects (*e.g.* glasses) placed on top of other planar objects (*e.g.* books). To resolve ambiguities such as objects stacked on other planar objects (such as books), the previously mentioned segmentation steps are repeated by treating each additional horizontal planar structure on top of the $\mathtt{t}_i$ table as a table itself. This leads to the correct segmentation of any hierarchically arranged supporting planes, such as books or boxes already located on tables, etc. The top row of the figure presents: the raw acquired point cloud data from the stereo camera (left), the estimation of surface curvatures with low-high represented as red-yellow (middle), and finally the resultant surface normals at the points in the cloud (right). The bottom row shows the segmentation results (left), the classification of the object clusters using the CRF model and FPFH features (middle), and the triangular surface meshes used as input for the grasping analysis.

Besides annotating points with a local geometric class and reconstructing surfaces, the system should be able to differentiate between various objects as well. To do this, we extend the FPFH idea to a Global Fast Point Feature Histogram (GFPFH) descriptor representation.

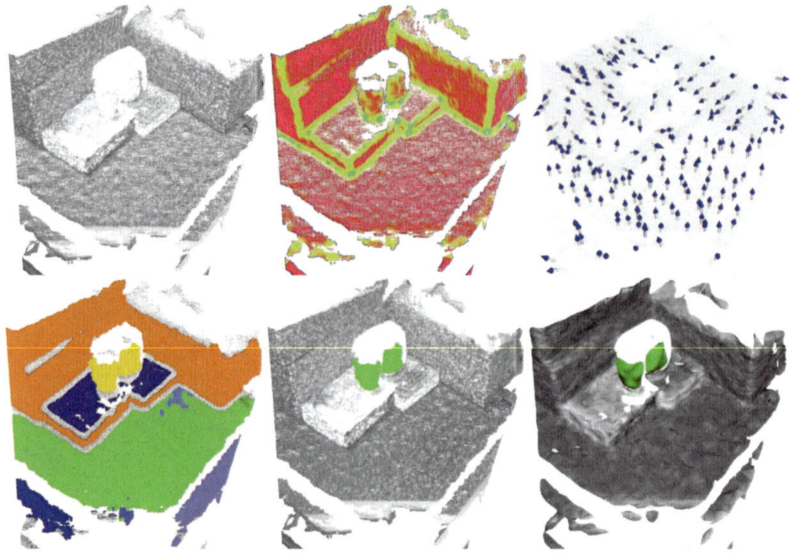

Fig. 8.25 An example of segmentation, classification, and surface reconstruction of complex scenes containing multiple objects located on top of other planar objects

The GFPFH computational steps are presented in Algorithm 8.1. The method assumes that for a set of input given points $\boldsymbol{p}_i \in \mathcal{P}$, a class label $\boldsymbol{c}_i \in \mathcal{C}$ has already been assigned through an a priori point based classification step using FPFH descriptors.

The algorithm begins by building an octree representation that spatially decomposes $\mathcal{P}$ into a set of *leaves* $\mathcal{L}$. Each leaf $\boldsymbol{l}_j \in \mathcal{L}$ contains a set of points $\mathcal{P}^l_j$ inside it, and due to the octree representation, each leaf is encapsulated into a parent leaf. This allows us to quickly create a multi-lod (multiple levels of detail) representation for the resultant GFPFH scheme.

A set of class probabilities $\mathsf{P}_{\boldsymbol{l}_i}(\boldsymbol{c}_j)$ are created for each leaf, where $\mathsf{P}_{\boldsymbol{l}_i}$ represents the probability of leaf $\boldsymbol{l}_i$ of being of class $\boldsymbol{c}_j$, and is estimated as:

$$\mathsf{P}_{\boldsymbol{l}_i}(\boldsymbol{c}_j) = \frac{\mathsf{n}_j}{\mathsf{n}} \cdot 100, \tag{8.4}$$

where n represents the total number of points in $\mathcal{P}^l_i$, and n_j represents the number of points having been estimated a class label j using the FPFH CRF model.

Then, for each pair of two leaves $\langle \boldsymbol{l}_i, \boldsymbol{l}_j \rangle$, a line segment r_{ij} is created, and a set of leaves $\mathcal{L}_{ij}$ intersecting with r_{ij} is obtained. Each leaf in $\mathcal{L}_{ij}$ is checked whether it is occupied or not (*i.e.* it contains points in it), and a histogram $\mathcal{H}$ is created for the pair, where:

$$\mathcal{H} = \mathsf{h}_{ij} = \begin{cases} 0, & \boldsymbol{l}_{ij} \text{ unoccupied} \\ \mathsf{P}_{\boldsymbol{l}_{ij}}, & \boldsymbol{l}_{ij} \text{ occupied} \end{cases} \tag{8.5}$$

Algorithm 8.1. The main computational steps for the Global Fast Point Feature Histogram descriptor

$\mathcal{P} = \{p_1 \cdots p_n\}$ *// set of 3D points*
$\mathcal{L} = \{l_1 \cdots l_m\}$ *// set of octree leaves encapsulating $\mathcal{P}$*
$\mathcal{C} = \{c_1 \cdots c_o\}$ *// set of point classes (generated from FPFH)*
for all $l_i \in \mathcal{L}$
  $\mathsf{P}_{l_i}(c_j)$ *// compute a list of probabilities P_{l_i} of l_i being of class c_j*
for all $l_i \in \mathcal{L}$
  // histogram holding the class probabilities of each leaf pair started from l_i
  $\mathcal{H} = \{\mathsf{h}_k | \mathsf{h}_k = 0\}$
  for all $l_j \in \mathcal{L}$
    $\mathsf{r}_{ij} \leftarrow (l_i, l_j)$ *// create a line segment r_{ij} between l_i and l_j*
    $\mathcal{L}_{ij} = \{l_{ij} | l_{ij} = \mathcal{L} \cap \mathsf{r}_{ij}\}$ *// get intersected leaves $\mathcal{L}_{ij}$ along the ray*
    if l_{ij} unoccupied *// is l_{ij} a free, unoccupied leaf in $\mathcal{L}$*
      $\mathsf{h}_{ij} = 0$
    else
      $\mathsf{h}_{ij} = \mathsf{P}_{l_{ij}}$
  // count the class changes for neighboring leaves in $\mathcal{H}$ and store in $\mathcal{H}_f$
  $\mathcal{H}_f \leftarrow \mathcal{H}$

This results in $(\mathsf{n}_l+1)\cdot \mathsf{n}_l/2$ histograms $\mathcal{H}$ of variable length for $\mathcal{L}$, one for each pair of leaves, where n_l is the number of leaves in $\mathcal{L}$. An example of such histogram is shown in Figure 8.26. For simplicity, instead of computing a probability for each class label in the leaf, we take the most dominant point class label as the leaf representative in the figure. The starting leaf l_i is represented with **S**, and its class is *4*, and the goal leaf l_j (represented with **G**) has class *3* as the most dominant one. Along the ray r_{ij}, three occupied leaves are intersected, with classes *4*, *3*, and *3*, followed by three unoccupied leaves.

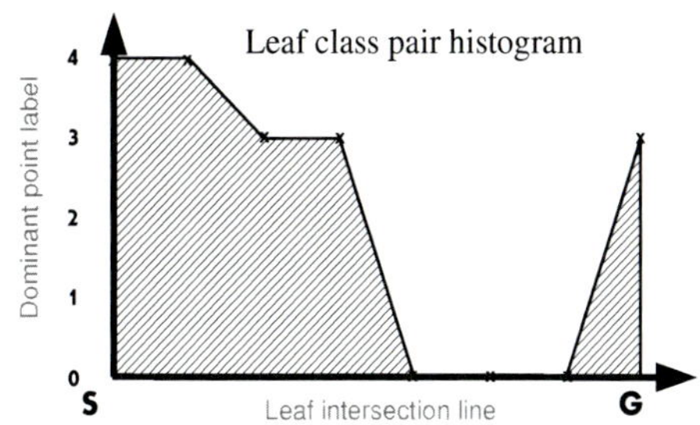

Fig. 8.26 An example of a resultant leaf class pair histogram, with the start leaf having the dominant class *4*, and the goal one *3*. Along the casted ray, three other occupied leaves are intersected, with classes *4*, *3*, and *3*, as well as three free leaves (represented with class *0* here).

Then, for each $\mathcal{H}_{ij}$ histogram, we estimate a fixed length histogram representation $\mathcal{H}_{f_{ij}}$. $\mathcal{H}_f$ consists of each possible combination of class neighbors, including the empty leaves, where the tuples $\langle \text{class}_a, \text{class}_b \rangle$ and $\langle \text{class}_b, \text{class}_a \rangle$ are equal, and is computed by counting the number of neighboring classes along the bins of $\mathcal{H}_{ij}$. Thus we get a set of fixed-length histogram representations of size:

$$\frac{(\mathsf{n}_c + 2) \cdot (\mathsf{n}_c + 1)}{2} \tag{8.6}$$

where n_c represents the number of FPFH classes.

The final GFPFH representation is obtained by computing a distance metric from each histogram in $\mathcal{H}_{f_{ij}}$ to the mean histogram of the set, and binning the values in a final histogram $\mathcal{G}$. Experiments with different metrics unveiled very good results using the Kullback-Leibler divergence:

$$d_k = \sum_{i=1}^{\frac{(\mathsf{n}_c+2)\cdot(\mathsf{n}_c+1)}{2}} (p_i^f - \mu_i) \cdot \ln \frac{p_i^f}{\mu_i} \tag{8.7}$$

where p_i^f and μ_i represent the $\mathcal{H}_f$ histogram value at bin i and the mean histogram of the entire set of $\mathcal{H}_f$ histograms at bin i respectively. Figure 8.27 presents some of the general theoretical aspects of the GFPFH estimation as presented in Algorithm 8.1.

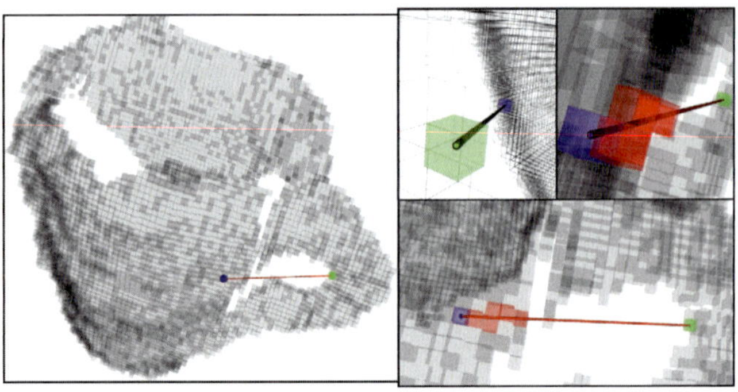

Fig. 8.27 The estimation of a GFPFH for a 3D point cluster. After a voxelized representation is created (left), for every two pairs of leaves, a ray is casted from the start leaf (green) to the goal one (blue). All intersections with other leaves and free space are recorded and ported into a leaf class pair histogram (see Figure 8.26).

To validate the proposed GFPFH scheme, a global descriptor was estimated for each segmented point cluster after all its points were annotated using one of the defined geometric classes. Each point cluster was spatially decomposed using a leaf size of 1cm, which led to an average of $\approx$ 300 leaves and a mean computation time

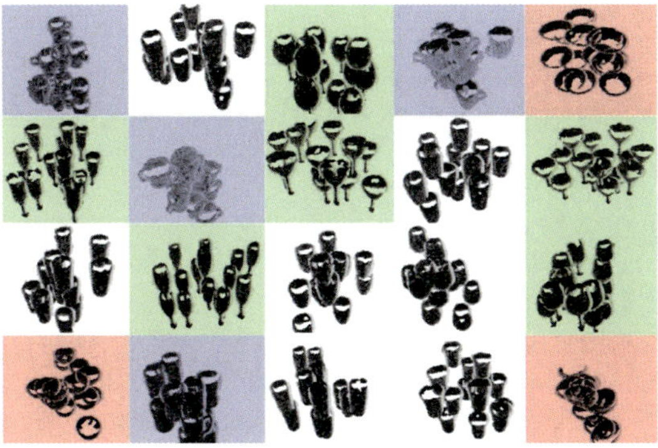

Fig. 8.28 The classification results for the objects in our dataset using the 4 devised GFPFH classes: i) simple glasses (white); ii) glasses with stems (green); iii) mugs with handles (blue); and iv) large bowls (red)

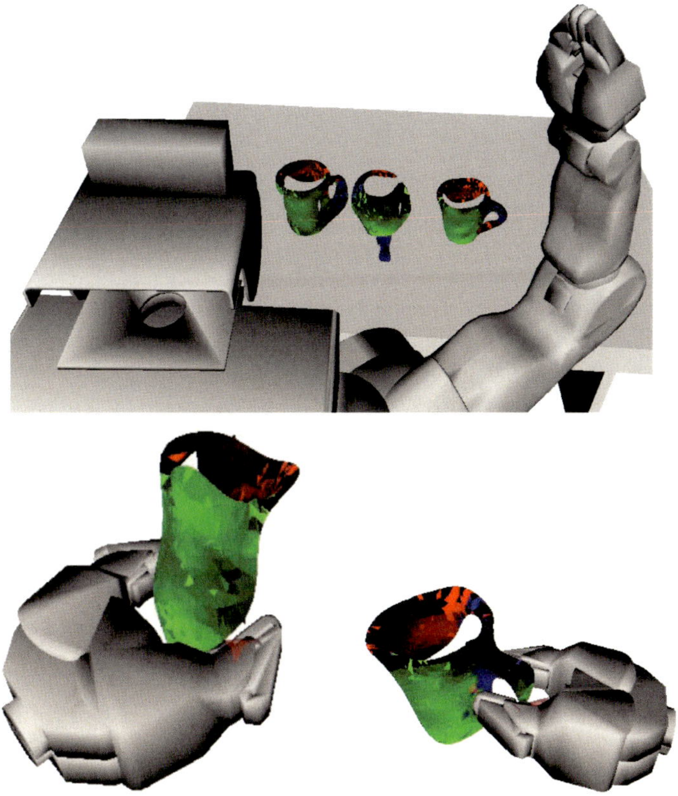

Fig. 8.29 Grasping scene examples acquired during simulation experiments performed using the PR2 [RSG$^+$09] robot model in OpenRAVE

of $\approx 2\,\text{s}$ per cluster. To train the model we selected a subset of 247 labeled objects, and devised four main categories of objects: i) glasses (no handles or stems); ii) glasses with stems (*e.g.* wine glass); iii) mugs (with handles) and finally iv) large bowls. The motivation of this class separation is that objects with very similar geometry are grasped in the same manner by our robot, and therefore it makes sense to group them together.

The object views were split into two equal parts, one for training the model and one for testing. The overall classification accuracy of the resultant SVM model was 95.13%. Figure 8.28 presents the classification of the 20 types of objects tested into the 4 devised categories.

The reconstructed surfaces can then be imported into a simulation tool for grasping analysis. An example of constrained grasp point estimation for a specific part of an object using the point labels in OpenRAVE [DK08] is shown in Figure 8.29. Using the GFPFH classification results (*i.e.* the object class) and the local surface labels acquired using the FPFH classification, the grasping of certain objects can be restricted to specific parts. For example, glasses with stems can be grasped only by using their stems, while for mugs the grasping estimation can be limited to the handle.

8.4 Summary

This chapter presented a set of novel scene interpretation mechanisms for point cloud data acquired using noisy sensors in indoor environments. The surface geometry around a point is characterized using multi-value histograms (PFH or FPFH), which provide pose invariant, discriminative feature representations even in the presence of varying density or noise. The proposed feature representations have shown to be useful for the problem of point classification with respect to the underlying geometric surface primitive the point is lying on. By carefully defining 3D shape primitives, and making use of machine learning classifiers, our framework can robustly segment point cloud scenes into geometric surface classes. For the problem of object categorization, a two layer classification system was proposed, first at a local point level for annotating geometric surface primitives, and the second at a global object level for grouping similar objects in the same class. The object classification is based on a novel global descriptor, which borrows from the theoretical properties of the local FPFH descriptor for point set cluster annotations: GFPFH.

Chapter 9
Parametric Shape Model Fitting

"A theory has only the alternative of being right or wrong. A model has a third possibility: it may be right, but irrelevant."

MANFRED EIGEN

This chapter tackles the problem of scene interpretation, including object reconstruction for everyday manipulation activities in domestic human living environments, in particular kitchens, out of point cloud data. The proposed framework takes raw point cloud datasets as input, and segments horizontal planar areas that support objects used in manipulation scenarios (see Figure 9.1).

Fig. 9.1 Segmenting a dataset representing a table scene and the object clusters located on it

The output is represented by a set of hybrid geometric object models that consist of shape coefficients and triangular surface meshes. The use of these hybrid models allow to synergetically combine the strengths of shape representations and those of surface models for a compact representations of objects. As Figure 9.2 illustrates, shape models enable the system to add surface information for parts of the object that were not covered by sensor data. Also, shape models are unique representations

R.B. Rusu: *Semantic 3D Object Maps for Everyday Robot Manipulation*, STAR 85, pp. 137–146.
DOI: 10.1007/978-3-642-35479-3_9

for which grasps can be determined in advance and be reparameterized on demand. In addition, exploiting the knowledge about the shape enables the robot to eliminate inaccuracies caused by sensor noise as it determines the best shape fit for the sensor data. However, the use of shape models assumes that objects have strong regularities, which is often but certainly not always satisfied by the objects of daily use in domestic human environments. Thus, the remaining parts of the object cluster are reconstructed through triangular meshes that can approximate arbitrary shapes better but have disadvantages such as: they are not compact, not generalizable, and yield higher geometric inaccuracies.

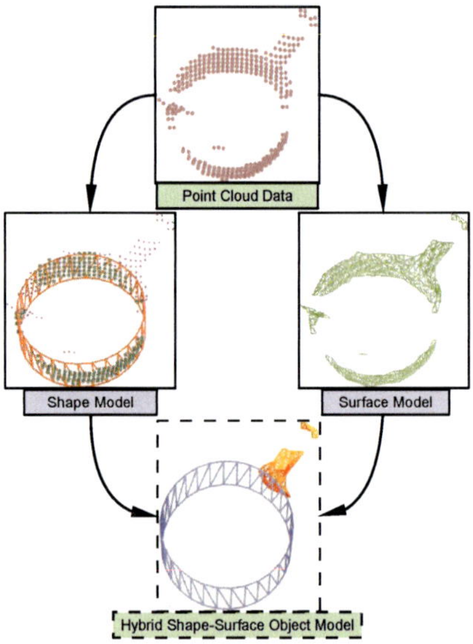

Fig. 9.2 The basic shape and surface model estimation step and the reconstruction of the hybrid shape-surface object model

Thus the proposed method first tries to explain parts of the point cloud through simple geometric shape models by fitting 3D geometric primitives such as spheres, cylinders, cones, and planes. It then reconstructs the remaining parts of the point cloud cluster using surface triangular meshes [MRB09], and combines the partial models of the point cloud cluster into a hierarchically structured hybrid model.

9.1 Object Segmentation

The architecture of the mapping system presented herein, is shown in Figure 9.3. Our methods take an unorganized 3D point cloud $\mathcal{P}$ as input and produce a set of shape coefficients $\mathcal{S}$ and a set of surface models $\mathcal{O}$ as output. We identify two main modules, namely the Scene Interpreter which extracts the supporting planes and

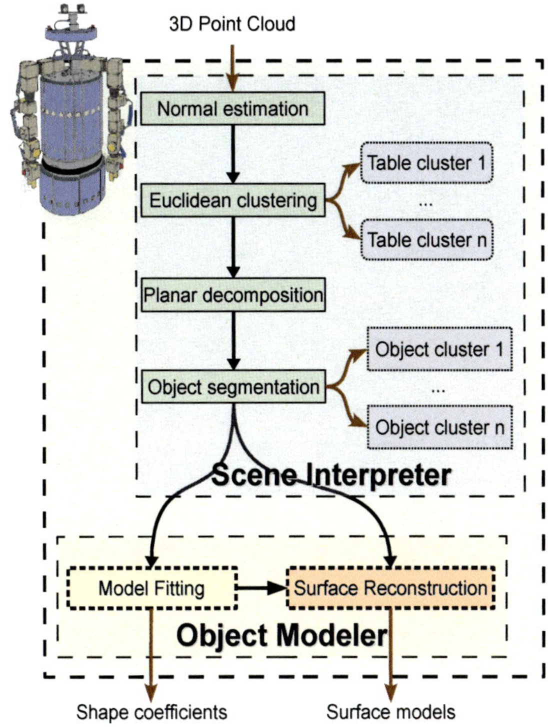

Fig. 9.3 The architecture of the proposed object reconstruction system. The input is comprised of unorganized 3D point cloud datasets, and the output is a set of reconstructed surface models $\mathcal{O}$ and a set of shape coefficients $\mathcal{S}$.

segments the object clusters, and the Object Modeler, which creates the shape and surface models for each segmented object cluster. To process the data faster, a world coordinate frame with the $\vec{z}$ axis pointing upwards is defined, and $\mathcal{P}$ is transformed into this frame.

The mapping pipeline can be described using the following computational steps:

1. For every 3D point $\boldsymbol{p}_i \in \mathcal{P}$, if no $\vec{\boldsymbol{n}}_i$ surface normal information is already present, an estimate of the normal to the best least-squares plane fit locally to the k-nearest neighbors of $\boldsymbol{p}_i$ is taken;
2. Point normals $\vec{\boldsymbol{n}}_i$ that are approximatively parallel with the world $\vec{z}$ axis, are grouped into a set $\mathcal{T} = \{\mathsf{t}_1 \cdots \mathsf{t}_n\}$ of Euclidean *table* candidate clusters;
3. For every cluster $\mathsf{t}_i \in \mathcal{T}$, a sample consensus robust estimator is used to find the best planar model;
4. For every set of point inliers $\mathcal{P}^i$ corresponding to the planar model for t_i found, a bounding polygon is computed as the table bounds;
5. All Euclidean point clusters $\mathcal{C}^i = \{\mathsf{c}_1^i \cdots \mathsf{c}_n^i\}$ supported by the table model (*i.e.* sitting on the table, and within the polygonal bounds) are extracted;

6. For every cluster $c_j^i \in \mathcal{C}^i$ a robust search for the best set of primitive shapes $\mathcal{S}^i$ is conducted;
7. A set of surface models $\mathcal{O}^i$ that takes $\mathcal{S}^i$ into account is built;
8. Finally the resultant hybrid shape-model $\mathcal{M}$ is built by concatenating the shape model with the surface model.

In general, the segmentation of planar areas out of sensed point cloud data representing indoor environments can be done in a bruteforce manner (as presented in [NH08] for example) by simply picking 3 random points, estimating the parameters of a plane from them, and then scoring points to this model using a distance threshold, in a sample consensus framework. While this approach needs no a-priori data processing and works in a straightforward manner on simple datasets, it might fail on more complex environments, where a resultant model might contain points from various different parts of a room, for example, located on different objects. Furthermore, a model could count as inliers points which do not respect the plane equation, that is, the estimated surface normal $\boldsymbol{n}_i$ at an inlier candidate point $\boldsymbol{p}_i$ is not parallel to the plane's normal $\vec{n}$.

To account for such complex environments, but also to considerably speed up the planar segmentation results, we proceed as follows. For a given point cloud dataset $\mathcal{P}$, we construct a downsampled version of it, $\mathcal{P}_d$, where $\boldsymbol{p}_j \in \mathcal{P}_d$ represents the centroid of a set of points $\mathcal{P}_j = \{\boldsymbol{p}_i \in \mathcal{P}\}$ obtained using a fixed spatial decomposition of the data (e.g. kD-tree). Furthermore, the normal $\vec{\boldsymbol{n}}_j$ to the underlying surface represented by $\mathcal{P}_d$ is estimated by fitting a local plane to the set of points $\mathcal{P}_j$, and approximating $\boldsymbol{n}_j$ as the normal $\vec{n}$ of the plane.

Then, in general, the system uses similar sample consensus techniques as [NH08], but imposes an additional constraint on the sample selection step, that is, for every two pair of points $\boldsymbol{p}_i, \boldsymbol{p}_j$ (with their estimated surface normals $\vec{\boldsymbol{n}}_i, \vec{\boldsymbol{n}}_j$) in the three required chosen samples: $\vec{\boldsymbol{n}}_i \cdot \vec{\boldsymbol{n}}_j \approx 0$.

For the purpose of the experiments presented in this chapter however, we are mostly interested in the segmentation of tables as horizontal planes which can support objects on them. Therefore, another optimization can be introduced in the planar decomposition approach, that is, from the the entire set of points $\boldsymbol{p}_i \in \mathcal{P}_d$, select only those with their estimated surface normals $\vec{\boldsymbol{n}}_i$ approximatively parallel with the world $\vec{z}$ axis. After that, we perform a fast clustering of the selected points in an Euclidean sense and construct a set $\mathcal{T} = \{t_1 \cdots t_n\}$ of *table* candidate clusters. This has the advantage that since the clusters are independent with respect to each other, the subsequent planar segmentation step can be performed in parallel for more than one cluster, thus decreasing the computational requirements considerably.

The next major step in the proposed geometric processing pipeline is the object segmentation step. Given a set of validated t_i models, a search for sets $\mathcal{C}^i$ of object candidates which are supported by these models is performed. These objects are said to be *movable*, in the sense that we are expecting the robot to be able to manipulate them. In a kitchen scenario they usually fall into the category of small kitchen utensils, dishware, food products, and so on.

To segment the set of objects, the t_i inliers are taken and a bounding 2D polygon is created for each candidate. Then, we look at the points whose projection on the t_i model falls inside the polygon. Because we treat the bounding as a convex problem, there are situations where the resulted shape includes additional points which do not lie on the table surface, as shown in Figure 9.4. To solve this, all the point candidates are split into regions in an Euclidean sense using an octree connectivity criterion (*i.e.* occupied neighboring leaves belong to one region), and then an additional constraint is imposed: each region's projection onto the t_i table has to have a minimum footprint. More specifically, the octree leaves of each point region are projected onto the octree leaves of the table candidate, and the leaf intersections are counted. If the footprint area is too small, the point region will not be considered, as it might simply be sensor noise. The result of these processing steps is a set of Euclidean point clusters $\mathcal{C}^i = \{c_1^i \cdots c_n^i\}$.

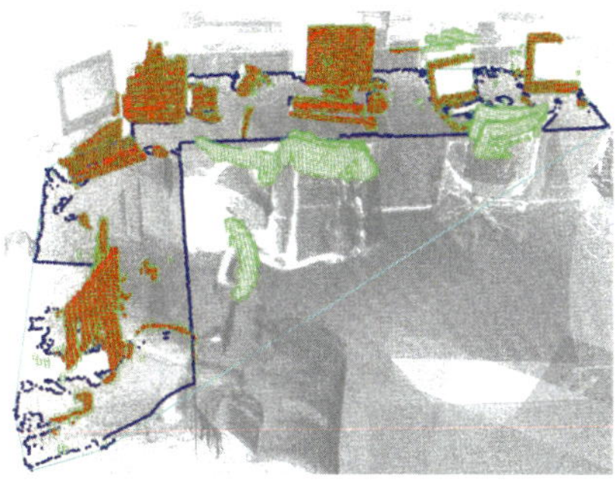

Fig. 9.4 Removing object candidates which fall within the convex polygon but do not intersect with the table. The table's boundary points are shown in blue, the convex hull lines with cyan, octree leaves with green, and the accepted object candidates in red.

Given the point cloud dataset in the left part of Figure 9.5, the remaining point clusters $\mathcal{C}^i$ after the previously mentioned segmentation steps are shown in the right part of the figure. Figure 9.6 presents more segmentation examples for various table scenes acquired in indoor environments.

9.2 Hybrid Shape-Surface Object Models

For each segmented point cluster c_i, a MSAC [TZ00] (M-Estimator Sample Consensus) based search is performed to find 3D primitive geometric surfaces such as planes, cylinders, spheres and cones.

The system maintains a list of all models $\mathcal{S}^i = \{s_1^i \cdots s_n^i\}$ that have been generated in the course of the search and scores each of them based on their number of inliers [RBMB09]. This list has the advantage that if a model is good enough to be accepted, the rest of the models get re-scored quickly by subtracting the inliers of the extracted model from their inlier lists. This way, the remaining models will be

Fig. 9.5 Left: A snapshot of our kitchen lab dataset, comprising roughly 15 millions of points. Right: Point clusters supported by tables or shelves (shown in yellow) which constitute candidates for movable object clusters (shown in random colors).

Fig. 9.6 Object cluster segmentation results for four different table setting scenes. The segmented table plane is marked with green, while the individual object clusters have a randomly assigned color.

found much faster, since we may have sampled them already. The implementations of these different models to be fitted share a common interface, so they can be easily enabled and disabled for different scenarios and new models can easily be added.

To achieve fast results, an octree is used for a localized sampling strategy similar to the one proposed in [SWK07a]. This results in an optimized inlier calculation by hierarchically pruning octree leaves that cannot contain inlier points to the current model. In addition, during inlier selection, points with normals contradicting those of the underlying model are discarded, and a connectivity criterion is enforced, *i.e.* we select the biggest connected component among the inlier candidates.

For each model s_i found, we perform a non-linear optimization of the shape parameters using the Levenberg-Marquardt algorithm [Sha98]. For some shapes, like cones, it proved beneficial to use a more restrictive distance threshold in the first inlier selection stage, and adjust this threshold to the expected noise level during the final inlier search. While this reduces the number of false positives and increases the number of false negatives, the refitting step converges to a better solution.

After extracting a shape, a search is performed for possible object *handles* by inspecting the vicinity of the model for fairly large clusters of points that can be grown from the current shape surface. These points get marked as handles to the current shape and are disregarded in the remaining model candidate list. This model-less approach was found to work more reliably than trying to fit specific geometric shapes like tori, since handle geometry can vary greatly. An example for a synthetic scene is given in Figure 9.7. Note that only the point positions are synthetically generated, as point normals are estimated, and no information on the original shape parameters is available to the object detection.

Fig. 9.7 A synthetic scene demonstrating the different estimated primitive shape models. Each primitive shape is marked with a different (random) color.

The left part of Figure 9.8 presents the results of cylinder and plane fitting on the point clusters from Figure 9.5. We noticed that these two models appear the most often in our kitchen datasets. The different cylinders are marked with different shades of blue, and the planes with shades of red. Notice that the cylinder radius was not constrained to a specific range, and therefore several small edges appear as cylinders due to the properties of the normal estimation algorithm. However, since the model parameters are known, these shapes can be grouped together and removed if necessary.

Our triangulation method [MRB09] propagates an advancing front of triangles, adding new triangles at each vertex by local projections of the nearby points onto a supporting plane. The right part of Figure 9.8 presents an example of surface models created for some of the objects presented in the left part of the figure. The resultant hybrid model is then created by triangulating the remaining outliers from the model fitting step, and adding it to the shape coefficients model as follows:

$$\mathcal{M} = \mathcal{S} \cup \mathcal{O} \tag{9.1}$$

To test the validity of the proposed approach, the modeling pipeline was applied to multiple datasets representing table setting scenes, acquired in a kitchen laboratory.

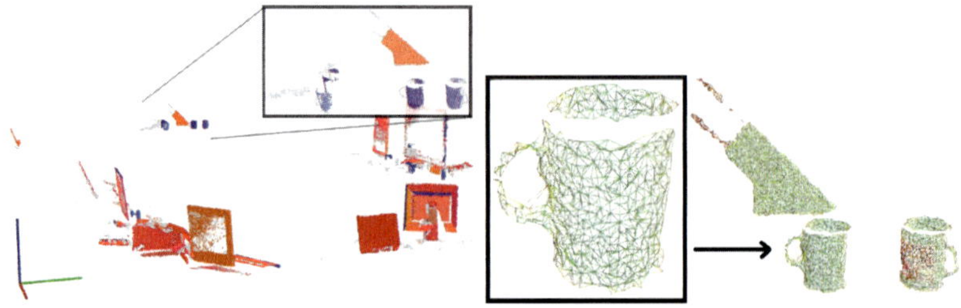

Fig. 9.8 Left: cylindrical (blue) and planar (red) shape candidates detected on tables. All remaining points are shown in gray. Right: examples of triangulated surface models for the kitchen counter point clusters.

Table 9.1 Model fitting results for four different table setting scenes. From top to bottom: segmented object clusters, shape models, surface models, and finally the hybrid shape-surface models.

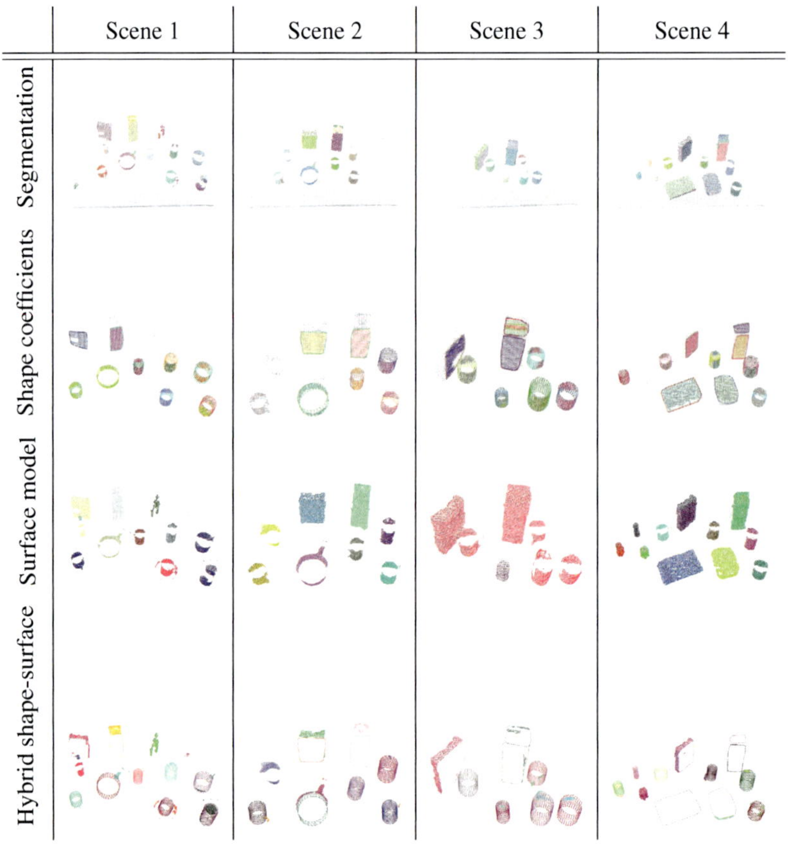

Fig. 9.9 Shape classification and fitting results for a cluttered scene without a priori object clustering: cylinders in blue, planes in red, and the remaining points in gray

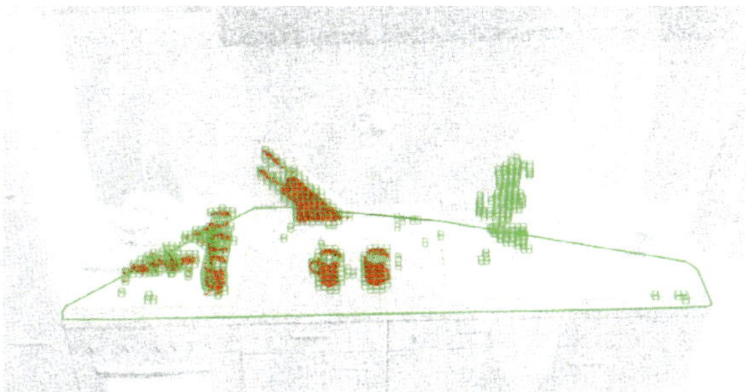

Fig. 9.10 Object clustering based on octree connectivity segmentation on a planar horizontal surface (*i.e.* table). Only points in red are selected as object candidates, as the rest do not fulfill the object footprint criterion.

Each of the computational steps previously described was ran on every scene, and the results were inspected by comparing the estimated shape models with real world measurements. Table 9.1 presents the results obtained for four different datasets. From top to bottom, the table presents: i) the object cluster segmentation from a raw point cloud dataset; ii) the shape model segmentation; iii) the surface models obtained by triangulating the points; and finally iv) the hybrid shape-surface models.

Even though the above results were obtained by applying our segmentation and model fitting methods on individual point clusters, the algorithms work directly for cluttered scenes where clustering is not possible due to all object surfaces being connected. Figure 9.9 shows the shape classification results for a cluttered table scene. Most shapes are reliably detected even if partially occluded.

Figure 9.10 presents an important limitation of the proposed mapping approach. Due to the fact that our sensing devices cannot return any measurements from shiny black surfaces such as the stove cooktop or the kitchen sink, it is impossible to determine the fact that there are any planar areas which could support objects there. Therefore, the horizontal table planar support is incomplete (as seen in Figure 9.10), and some points which could be part of some objects will be left out, as their footprint (support) on the table is very small or inexistent.

9.3 Summary

This chapter presented a comprehensive system for the acquisition of 3D hybrid shape-surface geometric models in close-range scenes useful for mobile manipulation and grasping. The models are acquired out of sensed point cloud data using robust shape segmentation and surface triangulation methods. By fitting primitive geometric shapes to the data, the proposed framework is able to reconstruct and infer missing data, and thus improve the resultant models for grasping. Furthermore, by splitting the object data into clusters, a decoupled triangular mesh map is created, which holds favorable computational updates in the presence of environment dynamics.

Part III
Applications

Chapter 10
Table Cleaning in Dynamic Environments

"Law of window cleaning: It's on the other side."

Personal robots are coming to age, enabling a multitude of problem scenarios that were unsolvable until recently, to be tackled. One of the remaining challenges however is to design systems that can function in the presence of humans without harming them, that is, in general operate safely in the presence of environment dynamics. Designing such a system might seem trivial, but for a platform capable of performing online without user interaction, it raises a few issues such as: i) how is the goal reached when the state of the environment changes? or ii) can real-time performance still be achieved if re-planning the robot's motion is necessary?

This chapter describes a system architecture that tries to identify and solve exactly these issues. The overall goal is to realize a basic perception for manipulation layer for a personal robot performing pick-and-place tasks in environments where humans live and work. This translates into the development of a framework that enables a mobile manipulation system to enter a previously unknown room and perform a certain task, like cleaning a table, in the presence of a human person. The personal robotics platform used to demonstrate the efficiency of this system architecture is the PR2 (Personal Robot 2) mobile manipulation platform from Willow Garage (see Figure 10.1).

For the problem of cleaning a table, the space of all possible objects must obviously be restricted to rigid everyday objects that are easy to grasp, such as cups, soda cans, tea boxes, bowls, etc. The important aspect is that these do not have to be known to the robot beforehand, which gives a certain degree of scalability to the system. This seemingly trivial manipulation task poses serious challenges for the state of the art in autonomous mobile manipulation. These challenges include the perception of object constellations on the table, the geometric reconstruction of the unknown objects to be picked up from partial sensor views, and the treatment of the rest of the scene as dynamic obstacles – in particular the person reaching into the operating space of the robot. The motion planning challenges include the seamless execution of planned motions when reaching for objects in the presence of moving obstacles, and the determination of appropriate grasps based on a partial and possibly inaccurate geometric reconstruction of the objects to be picked up.

R.B. Rusu: *Semantic 3D Object Maps for Everyday Robot Manipulation*, STAR 85, pp. 149–159.
DOI: 10.1007/978-3-642-35479-3_10

Fig. 10.1 The PR2 mobile robot platform used for the table cleaning experiments

There are not many manipulation systems that are capable of addressing these problems, due the requirements of successful operation of their many complex components, such as 3D perception, motion planning and more. With the exception of [SFW+08, BOW+07, ARA+06, KHY+06, QBN07, NJAK08], few other approaches have presented similar capabilities of functioning robustly and reliably over extended periods of time on real hardware. Due to their complexity, complete systems have not been fully realized yet, and the research problem remains open. The work presented here describes a system with real-time perception and manipulation capabilities, with an emphasis on robustness and software re-usability.

The proposed system architecture is presented in Figure 10.2. The hardware architecture of the PR2 platform is described in more detail in [RSG+09]. From a software implementation point of view, all the components necessary to perform the tasks presented here are modular so they can be reused individually, and are available as part of the ROS (Robot Operating System) open source initiative (see Chapter 3). Because of their modularity, they can be used for robots other than the PR2 as long as a respective robot description is available.

The software architecture presented in Figure 10.2, shows the complete system used for the purpose of the experiments, but with a focus on the components depicted in color: 3D perception and motion planning. The architecture can be viewed as a 3-tiered one. The lowest layer provides the PR2 hardware controllers, the middle layer contains the functional modules, and the top layer is represented by the task executive that manages the processes at the lower layers. In particular, the task executive parameterizes and synchronizes processes such as perception, navigation, reaching and grasping. Its purpose is to also provide context-specific control, responsiveness to sensory events, and failure detection, analysis, and recovery.

To be able to interact with its environment, a robotic system must first be able to perceive it, and it must do so accurately by detecting pertinent changes as they occur. This perception component is fundamental to both localization and motion planning. The 3D perception system consists of two main nodes which build two

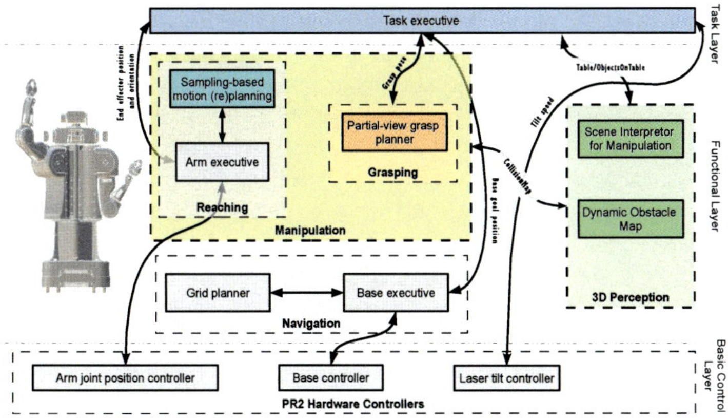

Fig. 10.2 The overall ROS system architecture. Boxes with continuous lines define nodes. Boxes with dashed lines delimit groups of nodes with related functionality. Communication between nodes is represented by arrows.

separate maps: a dynamic obstacle map used for collision detection in 3D while performing arm movements, and a scene interpreter which aggregates the acquired point cloud dataset with semantic annotations (e.g., class labels such as: floor, walls, tables, objects on table [RMB$^+$08a]). The input to the 3D perception module is a set of 2D laser scan messages, acquired from the tilting Hokuyo laser sensor installed on the head of the PR2 robot. Figure 10.3 presents a detailed version of the 3D perception pipeline used.

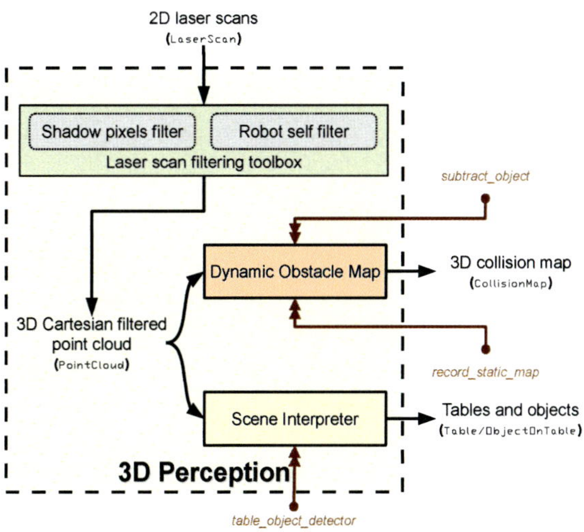

Fig. 10.3 Detailed diagram of the ROS 3D perception pipeline presented in Figure 10.2. The double arrow messages represent requests served by the nodes.

The sampling based motion re-planning module[CLH$^+$05, LaV06] requires a 3D map for collision avoidance (for example when moving one of the arms). From an architectural standpoint, sampling-based motion planning relies on a set of libraries with a modular design, keeping the actual motion planner, collision checking and robot modeling separate. The libraries are used together in a ROS node that monitors the robot state and receives updates from the dynamic obstacle map.

Before projecting each 2D laser scan into a 3D point cloud Cartesian frame, a set of preprocessing filters are applied, in order to clean the data of spurious measurements, as well as to remove points from the scan which are sampled from the robot body. The latter is achieved by treating the robot's body parts as convex hulls and checking the intersection of the laser scans with the hulls, while the shadow pixels filter removes erroneous points by comparing the angles formed between the lines created by subsequent scan measurements with the viewpoint, against a statistical threshold. After the data is partially cleaned, all the points left are projected into a global 3D coordinate frame. This constitutes the input of the two main nodes: the Dynamic Obstacle Map, and the Scene Interpreter.

10.1 Real-Time Collision Maps for Motion Re-planning

The Dynamic Obstacle Map is comprised of a set of geometric primitives created from the point cloud $\mathcal{P}$. This representation is updated in real-time for every new set of points, and its main usage is to provide support for collision avoidance (*i.e.*, reactive motion execution) for motion planning.

The node implementing the Dynamic Obstacle Map takes any set of points $\mathcal{P}$ that is available from the sensor, and inserts it into the map, which can generally be represented in multiple formats using: spheres, boxes, points, or triangle meshes. The latter is not recommended for situations where the data is extremely sparse, as incremental model updates become more difficult. In general, the node discretizes the space into smaller, uniform parts called *leaves*, with a certain user-defined resolution and publishes them in one of the above mentioned formats on the ROS network to the collision checker.

If the robot is to be allowed to operate and manipulate at high speeds, the laser tilt unit must be actuated as fast as possible, to account for the dynamics present in the environment. This however has the undesired effect that the total number of points per sweep is diminished considerably, and small or fast moving obstacles will not be present in the resultant map. To compensate for these types of situations, the node keeps a temporal sliding window (*i.e.*, data queue) for all data received, and computes the final obstacle map by integrating all the data together.

A deciding factor on the average computation time per map is represented by the number of data leaves that are requested from the motion planner, that is, the resolution of the world. Because there is no good fixed resolution that best accounts for all possible scenarios, the Dynamic Obstacle Map architecture is designed to be parameterizable on the fly. Furthermore, the task executive can specify what the space of interest around the robot is – the space that needs to be accounted for in the map. Taken together, these two constraints lead to significant computational decreases in the creation of the map. Algorithm 10.1 outlines the basic steps that are used to construct the map in real-time.

Algorithm 10.1. The main computational steps for the creation of the Dynamic Obstacle Map

$\mathsf{b}_r, \mathsf{r}_r$ *// robot bounds and resolution*
$\mathcal{P} = \{\boldsymbol{p}_1 \cdots \boldsymbol{p}_n\}$ *// set of 3D points*
$t_{\min} \stackrel{+}{=} 1, t_{\max} \stackrel{+}{=} 1$ *// increment sliding window time*
if $\neg(\exists \mathcal{L})$ *// no spatial decomposition exists (first $\mathcal{P}$)*
  $\mathcal{L} \leftarrow \mathsf{F}(\mathsf{b}_r, \mathsf{r}_r)$ *// create an empty list $\mathcal{L}$ of 3D leaves*
for all $\boldsymbol{p}_i \in \mathcal{P}$
  if $(-\mathsf{b}_r \leq \boldsymbol{p}_i \leq \mathsf{b}_r)$ *// check if inside bounds*
    estimate $(\boldsymbol{l}_{xyz} \in \mathcal{L}, \boldsymbol{p}_i \subset \boldsymbol{l}_{xyz})$ *// find the right leaf $\boldsymbol{l}$ for $\boldsymbol{p}_i$*
    add $\boldsymbol{p}_i$ to $\boldsymbol{l}_{xyz}$
for all $\boldsymbol{l}_{xyz} \in \mathcal{L}$
  if $(\boldsymbol{p}_j^t \leq t_{\min} \vee \boldsymbol{p}_j^t \geq t_{\max}, \boldsymbol{p}_j^t \subset \boldsymbol{l}_{xyz})$ *// check if outside sliding window*
    remove $\boldsymbol{p}_j$ from $\boldsymbol{l}_{xyz}$
$\mathcal{M} \leftarrow \mathsf{F}(\mathcal{L})$ *// create the final map from the set of leaves*

In addition to the periodical map creation, the node supports two important service calls, underlined with a double arrow in Figure 10.3. The *record_static_map* service call triggers a special mode in the node which records a complete sweep map as a static fixed reference map, that all recorded data will be merged with until the next service call. This translates into:

$$\mathcal{M}_f = \mathcal{M} \cup \mathcal{M}_r, \tag{10.1}$$

where $\mathcal{M}$ represents the map created from the current set of leaves, $\mathcal{M}_r$ is the static fixed reference map recorded at the service call, and $\mathcal{M}_f$ is the final 3D dynamic obstacle map.

Having a static reference map is beneficial during longer manipulation scenarios, where the estimated collision map may contain holes due to occlusions caused by the robot's arms at certain positions in time. This may in turn negatively affect the motion planning algorithms, which in the absence of pertinent solutions will attempt to use this occluded space, even though it has a certain probability of still being occupied. By recording reference maps and combining them, this problem can be alleviated. Furthermore, the Scene Interpreter can decompose regions of interest in the reference map such as the table for example, and remove everything else (such as the objects sitting on top of it), as shown in Section 10.2. Thus, the reference maps can be updated only with those parts of the environment which have a higher probability of being fixed over subsequent pick and place operations.

The second service call, *subtract_object*, introduces a negative collision map created from a given object that can be continuously subtracted from the computed maps, that is:

$$\mathcal{M}_f = \mathcal{M} \setminus \mathcal{M}_o, \tag{10.2}$$

where $\mathcal{M}_o$ represents the negative map created from the given object. The object is given as either a list of leaves or an oriented 3D bounding box. A special use case of this service call is to subtract objects from the collision map currently being manipulated with the gripper (see Figure 10.4). To achieve this, the node subscribes to the ROS network and inquires for the current position of the end-effector, and uses the given oriented object bounds to subtract the data. Please note that the object present in the gripper from the left part of Figure 10.4 is incomplete in $\mathcal{P}$, because parts of it are removed by the robot self filter at the laser scan level, due to the convex hulls of the gripper being much larger than the actual gripper itself.

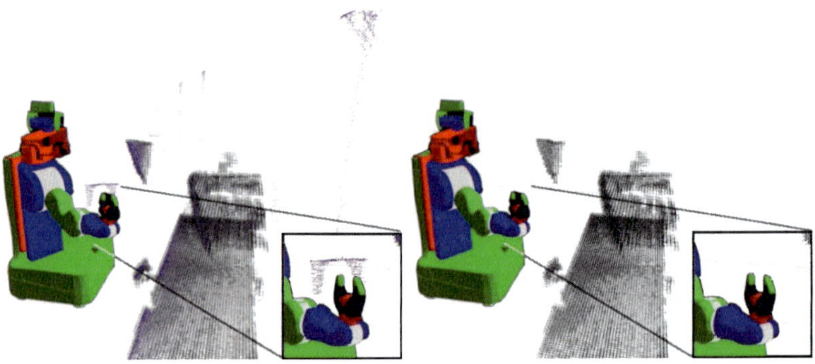

Fig. 10.4 Left: a dynamic obstacle map representation over-imposed on the point cloud dataset; right: the dynamic obstacle map alone. Please notice that the object held in the gripper is part of the point cloud, but has been removed using the subtract object service call from the map.

10.2 Semantic Interpretation of 3D Point Cloud Maps

The goal of the second component of the 3D perception pipeline, the Scene Interpreter node, is to continuously segment and label surfaces in the world with semantic annotations. As previously presented in Chapters 8 and 9, these annotations can be acquired either by using 3D features and machine learning classifiers, or by directly performing geometric model fitting in the input data.

For the purpose of the experiments performed herein, the most important areas for pick and place operations are tables and other horizontal planes which can support objects. Therefore there is no need to create a feature set and train a model, as it suffices to construct a single heuristic rule for determining all major horizontal planar areas in the world that the robot could operate on. The vertical bounds of the search are given by the physical constraints of the robot arms, *i.e.*, what is the lowest and highest place where the robot could pick an object from. Figure 10.5 presents a more general scene interpretation for planar areas segmented from a $\mathcal{P}$ point cloud dataset. The class labels identified are: floor, ceiling, walls (vertical large structures), and tables (or better said: horizontal planes which can support objects sitting on them). All remaining points are uniformly labeled as obstacles.

Algorithm 10.2. The main computational steps for determining the list of tables and objects located on them

$\mathsf{b}^z_{\min}, \mathsf{b}^z_{\max}$ *// min/max reachable arm positions on z, i.e., table bounds*
$\mathcal{P} = \{\boldsymbol{p}_1 \cdots \boldsymbol{p}_n\}$ *// set of 3D points*
$\mathcal{P}_b = \{\boldsymbol{p}_i, \mathsf{b}^z_{\min} \leq \boldsymbol{p}^z_i \leq \mathsf{b}^z_{\max}\}$ *// subset of all 3D points within table bounds*
for all $\boldsymbol{p}_i \in \mathcal{P}_b$
  estimate $(\vec{\boldsymbol{n}}_i \text{ from } \mathcal{P}^k)$ *// estimate surface normal from nearest neighbors*
  if $(\alpha = \vec{\boldsymbol{n}}_i \times \vec{\mathsf{z}} \approx 0)$ *// check if the normal is parallel to the $\vec{\mathsf{z}}$ axis*
    $\mathcal{P}_z \leftarrow \boldsymbol{p}_i$ *// add $\boldsymbol{p}_i$ to the $\mathcal{P}_z$ set*
estimate $(\mathcal{C} = \{\mathcal{P}^1_z \cdots \mathcal{P}^n_z\}, \mathcal{P}^i_z \subset \mathcal{P}_z)$ *// break $\mathcal{P}_z$ into Euclidean clusters*
for all $\mathsf{c}_i = \mathcal{P}^i_z \in \mathcal{C}$
  // find the best plane fit using sample consensus
  estimate $(\{\boldsymbol{a}, \boldsymbol{b}, \boldsymbol{c}, \boldsymbol{d}\}, \boldsymbol{a} \cdot \boldsymbol{p}^x_i + \boldsymbol{b} \cdot \boldsymbol{p}^y_i + \boldsymbol{c} \cdot \boldsymbol{p}^z_i + \boldsymbol{d} = 0, \boldsymbol{p}_i \in \mathsf{c}_i)$
  estimate $(a_{\min}, a_{\max})$ *// find the min/max bounds of the planar area*
  $\mathcal{M} \leftarrow F(\mathsf{c}_i)$ *// add the table parameters to the final map*
  for all $\boldsymbol{p}_i \in \mathcal{P}$
    if $(a^x_{\min} \leq \boldsymbol{n}^x_i \leq a^y_{\max}, a^y_{\min} \leq \boldsymbol{n}^y_i \leq a^y_{\max})$ *// within bounds?*
      $\mathcal{P}_o \leftarrow \boldsymbol{p}_i$ *// add $\boldsymbol{p}_i$ to the $\mathcal{P}_o$ set*
  estimate $(\mathcal{O} = \{\mathcal{P}^1_o \cdots \mathcal{P}^n_o\}, \mathcal{P}^i_o \subset \mathcal{P}_o)$ *// break $\mathcal{P}_o$ into Euclidean clusters*
  for all $\mathsf{o}_i \in \mathcal{O}$
    $\mathcal{M} \leftarrow \mathsf{F}(\mathsf{o}_i)$ *// add the object parameters to the final map*

From an implementation point of view, the node has two operation modes:

- a continuous mode where for each acquired set of points $\mathcal{P}$, it labels all major planar areas;
- a service mode, where the task executive requests a list of tables in view and objects sitting on them.

Algorithm 10.2 presents a brief description of the computational steps invoked for finding tables and objects on them when the node is running in the service mode, and partial results are presented in Figure 10.6.

The resultant map $\mathcal{M}$ contains a set of tables with their planar equations and bounds, and a set of object models $\mathcal{O} = \{\mathsf{o}_1 \cdots \mathsf{o}_n\}$. These object models represent compact partial-views representations for the objects supported by tables in the real world. Their usage is twofold: i) they provide the exact goal positions and object extents for the gripper; and ii) they support the *subtract_object* service call from the Dynamic Obstacle Map, once the object is picked up.

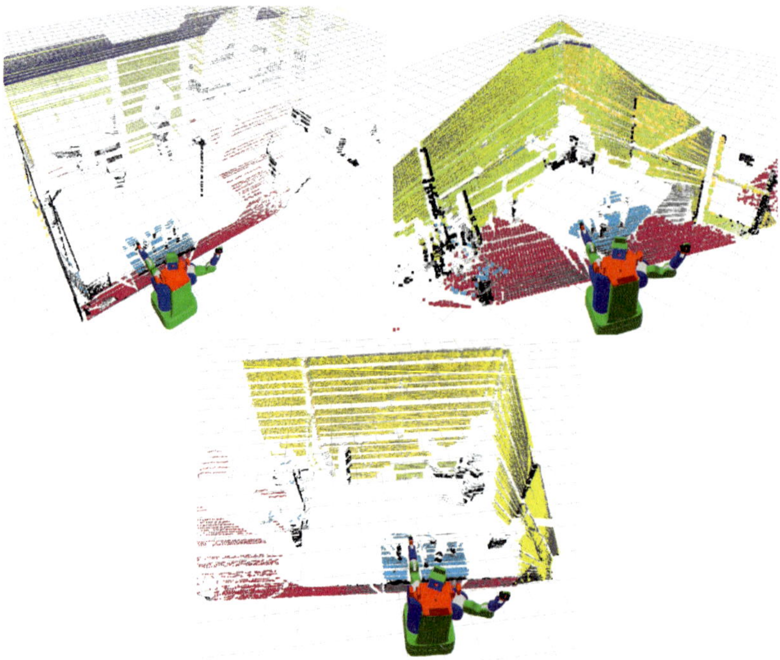

Fig. 10.5 Scene examples with point annotations labeled with the Scene Interpreter: floor (dark red), ceiling (dark blue), yellow (walls), and light blue (tables)

10.3 System Evaluation

The proposed architecture has been evaluated via multiple experiments on the PR2 robot, that validated the tight integration of each system component[1]. The sampling-based motion planner [RSG$^+$09] uses the dynamic obstacle map to compute a safe 3D path without colliding with the obstacles seen by the sensor. Its output is a kinematic path that takes the arm from the current position to the goal.

The partial view grasp planner is in principle a zeroth-order grasper. It uses the object models sent by the scene interpreter node, and assumes that a direct approach with the gripper in horizontal orientation will properly grasp an object. This limits the set of objects to be grasped to those which fall between the maximum opening bounds of the gripper.

To show the integration of all the individual components presented in Figure 10.3, we designed the following scenario:

1. The task executive triggers a slow laser scan and calls the Scene Interpreter's service request for a list of table candidates with objects on them;

[1] `http://www.willowgarage.com/iros2009-tabletop-manipulation` contains demonstration videos taken during the experiments.

Fig. 10.6 The extraction of a table surface and the individual object clusters supported by it, from a partial view

2. If the list of tables returned is not within reach, the task executive plans a 2D trajectory for the base, and moves the robot in the vicinity of the closest table;
3. The task executive triggers fast laser scans and starts building the Dynamic Obstacle Map;
4. From the list of objects on table, one is randomly chosen, and its goal pose is given to the motion planner, which computes a safe trajectory for the arm;
5. While the arm is moving, the information provided by the Dynamic Obstacle Map is used to re-plan the trajectory in case it's deemed as no longer safe;
6. The scenario ends when the end effector reaches the goal position and the object is in a graspable state.

An example of a path planned for the arm is shown in Figure 10.7. Figure 10.8 presents the time spent for the computation of the dynamic obstacle map, and the trajectory to follow respectively, for one such experiment. The four plots shown in each figure represent laser sweeps of: 1 second in a static scene (1s_static), 1 second in a dynamic scene (1s_dynamic), 2 seconds in a static scene (2s_static), and 2 seconds in a dynamic scene (2s_dynamic) respectively.

As shown, both for laser sweeps with a period of 1 or 2 seconds, all components manage to finish their computations in real-time, that is before the next state update. The perception system reliably detected a candidate table and the object clusters sitting on it, the only exception being environments where multiple tables were located approximatively at the same distance from the robot. The planning component had a success rate of 89%, with only seldom failures being observed due to inaccuracies in modeling our obstacles which sometimes caused the planner to consider the initial state of the robot was in collision. Figure 10.9 presents a few example snapshots taken during one of the experiments concerning motion re-planning using 3D collision maps.

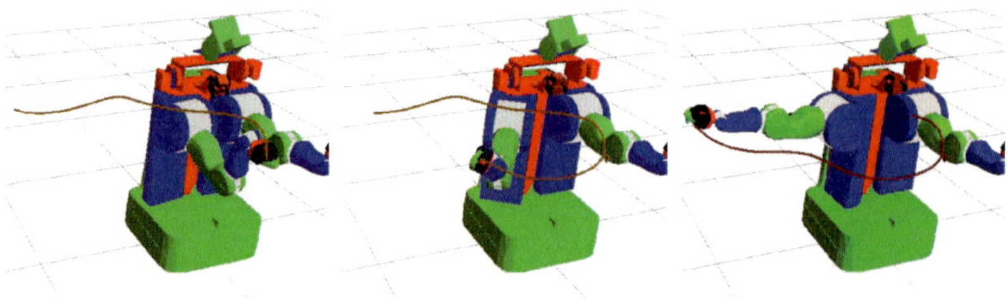

Fig. 10.7 Three instances of a path plan being executed in the context of dynamic obstacles, together with the trail of the end-effector

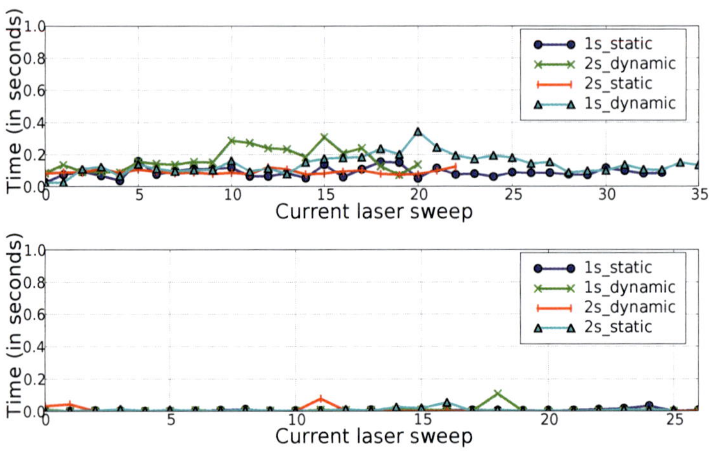

Fig. 10.8 Top: Dynamic Obstacle Map computation time plot (queue of 5 seconds), bottom: KPIECE [RSG$^+$09] motion planning computation plot

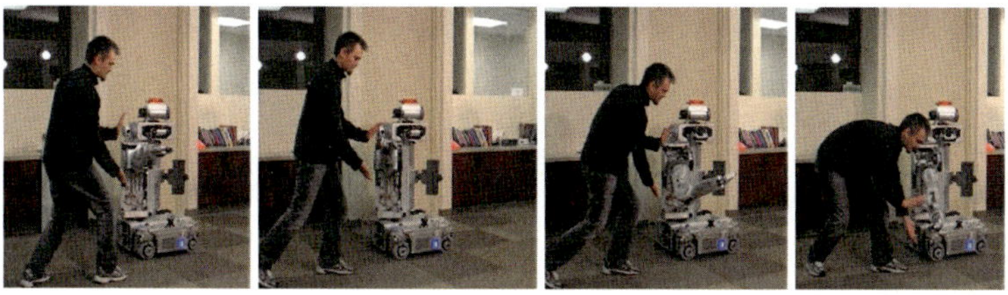

Fig. 10.9 Snapshots taken during one of the experimental trials concerning motion re-planning using 3D collision maps

10.4 Summary

This chapter presented a modular and distributed framework for a system that combines online 3D perception with mobile manipulation for personal robotics applications. To validate its efficiency, the integration of 3D collision maps and semantic annotations created in real-time from sensed laser data with a motion re-planning framework have been shown to work as building blocks towards efficient and robust pick and place scenarios, in the presence of environment dynamics.

Chapter 11
Identifying and Opening Doors

> *"Sometimes we stare so long at a door that is closing that we see too late the one that is open."*
>
> ALEXANDER G. BELL (1847 - 1922)

An important challenge for autonomous personal robots is to be able to enter a human living environment and function in it, that is, find ways of navigating and interacting with the world in an effective manner. This means the robot should be capable of going anywhere where it can physically fit, where it would be able to find energy sources and recharge its batteries when their capacity is running low, and in general it must be able to do useful things such as cleaning tables. These behaviors need the support of complex perception routines which can recognize power plugs, certain structures and objects in the world, etc. Some of these environment structures, such as fixtures (handles and knobs) on doors and pieces of furniture, are of key importance for the robot's performance. Any robot that will operate indoors must be able to correctly locate these fixtures and open doors to be able to better carry out different tasks. Since the robot may have to function in a wide variety of environments under varying lighting conditions, a robust door detection ability is essential for the robot.

The requirements of a perception system able to robustly identify doors and their handles are tied to the overall application goals. A number of systems have been proposed recently that are able to identify doors [AA07, AT08, ALAM06] and handles [EJL07, EK08, MK08, OBBH07] in 2D images. Though in general these approaches work reasonably well, they are all prone to the same problems, including illumination changes and a large number of false positives, to name a few. The same concerns apply to stereo-based systems, which in addition become unsuitable in environments where the lack of texture makes it difficult to recover the 3D structure of walls and doors.

Modeling the 3D characteristics of the door or the environment itself can be of great importance if the application includes mobile manipulation tasks such as opening the door, in addition to its detection. An architecture that renders reliable 3D maps of the world useful for collision detection for arm planning is therefore required. Also, the identification process should be sensitive to illumination changes

R.B. Rusu: *Semantic 3D Object Maps for Everyday Robot Manipulation*, STAR 85, pp. 161–175.
DOI: 10.1007/978-3-642-35479-3_11

as little as possible, and operate independently of the light sources present, possibly even in the dark. A feasible solution is to rely less on appearance and more on the actual geometry of the door and of the handle respectively.

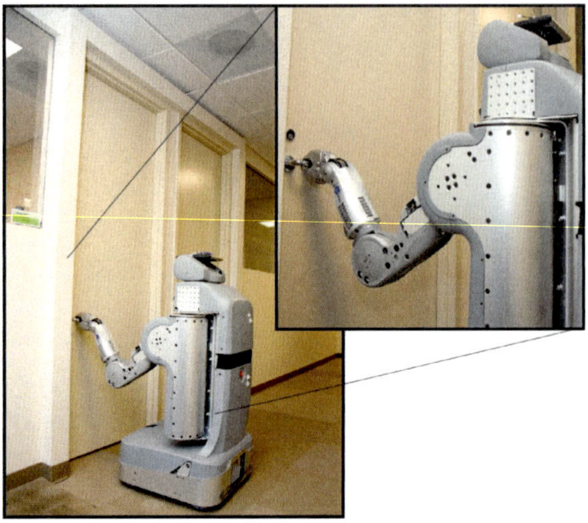

Fig. 11.1 Snapshot of the PR2 mobile robot during door identification and opening experiments

The system architecture presented in Figure 11.2 fulfills the above requirements. Instead of relying on computer vision techniques as other similar research initiatives, the system makes use of 3D point cloud data acquired using a sweeping 2D laser sensor to model the geometry of doors and handles. There are three principal subsystems, each dealing with a particular aspect of the mobile platform's intended operation, namely: 3D perception, navigation, and manipulation. All of them communicate with a set of hardware controllers and are controlled by a higher level task executive. To distribute and modularize the various components of the proposed architecture, each of the above mentioned subsystems is programmed as a collection of ROS (Robot Operating System) nodes (see Chapter 3 for more details). In particular, the 3D perception component includes a *door detection* and a *handle detection* node.

The experiments used for validating the proposed approach were carried out on a PR2 mobile manipulation platform at Willow Garage [RMCB09]. The PR2 (see Figure 11.1) is equipped with a variety of sensors, including a Videre stereo camera on a pan-tilt unit, and two Hokuyo UTM-30 laser range finders, one mounted on the mobile base, and one on a tilt stage (see Figure 11.3). The second Hokuyo is tilted up and down continuously, providing a 3D view of the area in front of the robot. The resultant point cloud, which contains both position and intensity information, is the main input of the perception system.

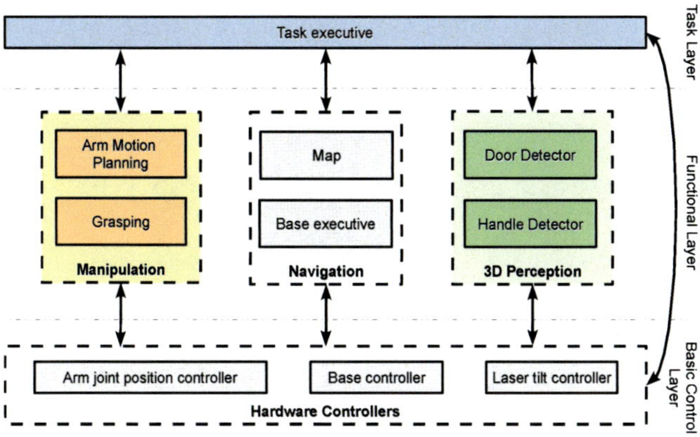

Fig. 11.2 The ROS system architecture for door identification and opening. Boxes with continuous lines define nodes. Boxes with dashed lines delimit groups of nodes with related functionality. Communication between nodes is represented by arrows.

The space of possible doors and handles is huge. In particular, handles come in a wide variety of shapes and sizes and could be mounted anywhere on a door. This makes the task of searching for doors and handles in a point cloud extremely difficult. A viable solution is to restrict the search space by considering only doors and handles that conform to the American Disability Act (ADA). The PR2 program aims to develop a manipulation platform that is ADA compliant, *i.e.*, the robot can access parts of the environment that a person in a wheelchair should be able to access. ADA compliance places constraints on the geometry and placement of handles, specifying in particular that handles must be above a certain height on the door and should be able to be opened without using a grasping hold. This simplifies the search for handles once a candidate door is found. It also simplifies the process of finding a point on the handle to grab, since the handles tend to be linear with a long lever arm rather than knob-like. ADA compliance also places a restriction on the width of the door since it must be at least wide enough to let a wheelchair through.

The proposed mapping approach builds on some of these constraints to achieve robust door and handle detection. The acquired point clouds are downsampled and annotated first to find candidate door planes. Several rules based on the constraints arising from ADA rules are then applied to prune the search area further and develop a *goodness* score for each candidate door. The candidate with the highest score represents the best choice for a door. A search within a limited area of the door plane is then conducted to find handles. The search is based on multiple criteria including the expected shape of the handle itself (linear) and the difference in intensities between the handles and the door plane. The segmentation method does not need any pre-determined thresholds, but instead automatically clusters and segments the regions of interest using a dual intensity and geometric analysis. Figure 11.4 presents an example of door and handle identification in a 3D point cloud dataset.

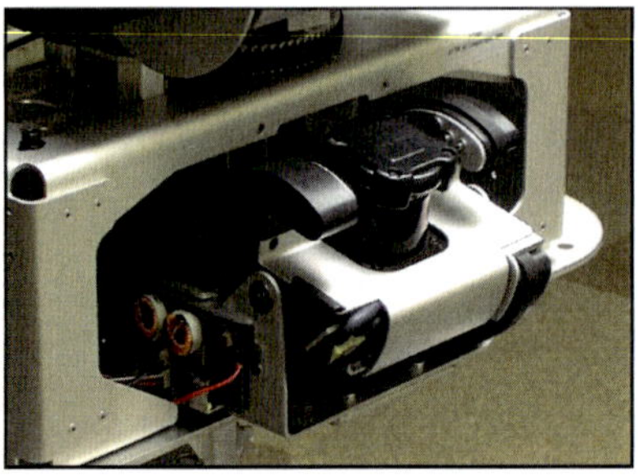

Fig. 11.3 The PR2 tilting unit platform with a mounted Hokuyo laser sensor

Fig. 11.4 Door and handle identification example in a 3D point cloud acquired using the tilting Hokuyo laser on the PR2. The handle is marked with green, the door frame with black, while the rest of the point cloud is shown in intensity (yellow-red shades).

11.1 Detecting Doors

The *door detection* node operates directly on the acquired point cloud data $\mathcal{P}$ from the tilting laser sensor, and makes no use of camera images or any other information sources to identify the door plane and its 3D bounds. The main motivation is given by the system's requirements to operate in a variety of situations, and account for any changes in the room ambient light throughout long periods of time. In addition, the perception system natively generates rich 3D annotations for the world map, using the geometric information captured in the point cloud, and is thus an invaluable source of information for the motion planner and grasping system, providing real-time map updates for collision detection (for the entire robot) and the desired 3D poses for the task goals (see Chapter 10).

The output of the node is represented by a list of doors described through a set of attributes including the door planes, their bounds, as well as their location with respect to a given world coordinate system. The doors are scored and returned with respect to a given fitness function $\mathcal{F}$, which includes parameters such as the number of point inliers supporting the door model or the current distance of the model to the robot. Any of the subsequent operations applied to a door candidate can uniformly be applied to the rest of the candidates, without any loss of generality.

Because the perception system is expected to run online, the computational time requirements for the door detection are restricted to the maximum allowable time difference between the acquisition of two subsequent point cloud datasets $\mathcal{P}_i$ and $\mathcal{P}_{i+1}$. This constitutes the only real time constraint of the application. Algorithm 11.1 presents the main computational steps that are used by the *door detection* node.

Algorithm 11.1. The main computational steps for the detection of doors

$\mathsf{b}_{\mathrm{ADA}}, \mathsf{b}_r, \mathsf{v}_p$ *// ADA door requirements, robot bounds, $\mathcal{P}$ acquisition viewpoint*
$\mathcal{P} = \{\boldsymbol{p}_1 \cdots \boldsymbol{p}_n\}$ *// set of 3D points*
$\mathcal{P}_d \leftarrow \mathsf{F}(\mathcal{P})$ *// create a downsampled representation $\mathcal{P}_d$*
$\mathcal{P}_d \leftarrow \{\vec{\boldsymbol{n}}_1 \cdots \vec{\boldsymbol{n}}_n \mid \vec{\boldsymbol{n}}_i \cdot (\mathsf{v}_p - \boldsymbol{p}_i) > 0\}$ *// estimate normals at $\boldsymbol{p}_i \in \mathcal{P}_d$*
$\mathcal{P}_{\vec{\mathsf{z}}} = \{\boldsymbol{p}_i \mid \vec{\boldsymbol{n}}_i \cdot \vec{\mathsf{z}} \approx 0\}$ *// select the set $\mathcal{P}_{\vec{\mathsf{z}}}$ with normals perpendicular to $\vec{\mathsf{z}}$*
estimate $(\mathcal{C} = \{\mathcal{P}^1_{\vec{\mathsf{z}}} \cdots \mathcal{P}^n_{\vec{\mathsf{z}}}\}, \mathcal{P}^i_{\vec{\mathsf{z}}} \subset \mathcal{P}_{\vec{\mathsf{z}}})$ *// break $\mathcal{P}_{\vec{\mathsf{z}}}$ into Euclidean clusters*
for all $\mathsf{c}_i = \mathcal{P}^i_{\mathsf{z}} \in \mathcal{C}$
  // find the best plane fit using sample consensus
  estimate $(\{\boldsymbol{a}, \boldsymbol{b}, \boldsymbol{c}, \boldsymbol{d}\}, \boldsymbol{a} \cdot \boldsymbol{p}_i^x + \boldsymbol{b} \cdot \boldsymbol{p}_i^y + \boldsymbol{c} \cdot \boldsymbol{p}_i^z + \boldsymbol{d} = 0, \boldsymbol{p}_i \in \mathsf{c}_i)$
  estimate $(\mathcal{A} = a_1 \cdots a_n)$ *// estimate geometric attributes for the planar area*
  if $\mathsf{F}(\mathsf{c}_i, \mathcal{A}, \mathsf{b}_{\mathrm{ADA}}, \mathsf{b}_r)$ *// does c_i respect the given constraints?*
    $\mathcal{D} \leftarrow \mathsf{c}_i$ *// add to $\mathcal{D}$, the list of good candidates*

The first step takes the set of input points $\mathcal{P}$, and creates a downsampled representation of them $\mathcal{P}_d$ using a fast octree structure. For each point $\boldsymbol{p}_i \in \mathcal{P}_d$, a surface normal $\vec{\boldsymbol{n}}_i$ is estimated by fitting a least-squares plane to the neighborhood $\mathcal{P}_i^k$ of $\boldsymbol{p}_i$. Due to the fact that the dataset is spatially sparser after downsampling, $\mathcal{P}_i^k$ is selected as the set of point neighbors from the original cloud $\mathcal{P}$. Given a viewpoint v_p from where the point cloud dataset was originally acquired, all resultant point normals $\vec{\boldsymbol{n}}_i$ must satisfy the equation:

$$\vec{\boldsymbol{n}}_i \cdot (\mathsf{v}_p - \boldsymbol{p}_i) > 0 \tag{11.1}$$

Then, $\mathcal{P}_d$ is transformed into a coordinate system defined at the base of the robot, with $\vec{\mathsf{z}}$ pointing upwards, and a subset of all the points $\mathcal{P}_{\vec{\mathsf{z}}}$ is selected, having their estimated surface normals $\vec{\boldsymbol{n}}_i$ approximatively perpendicular to the world $\vec{\mathsf{z}}$-axis, *i.e.*, $\vec{\boldsymbol{n}}_i \cdot \vec{\mathsf{z}} \approx 0$. The resultant set $\mathcal{P}_{\vec{\mathsf{z}}}$ is split into a set of Euclidean clusters using a region growing approach: $\mathcal{P}_{\vec{\mathsf{z}}} = \{\mathsf{c}_1 \cdots \mathsf{c}_n\}$. Each cluster c_i represents a potential door candidate in the framework, and a plane model is fit to each of them using a RMSAC (Randomized M-Estimator Sample Consensus) robust estimator [TZ00, CM02]. Due to the geometric independence of the clusters, the search is parallelized by dynamically selecting a set of N clusters for concurrent processing, based on the number of CPUs present in the system and their current availability. The inliers of the models found are then projected onto their respective planes and a set of bounding 2D polygonal structures is estimated.

Since an estimated door plane normal is perpendicular to the $\vec{\mathsf{z}}$ axis, a final robust estimation step is performed to fit the two best vertical lines in each aforementioned 2D polygon, and thus estimate the two major door edges. From these, a set of geometric attributes $\mathcal{A}$ is estimated, including: the width, height, minimum and maximum values along each axis, area, number of supporting inliers, and finally the height of the two door edges (parallel to $\vec{\mathsf{z}}$). These attributes constitute the input to the geometric tests used to select and weight the best door candidates $\mathcal{D}$ for a given dataset $\mathcal{P}$.

The tests refer mainly to the ADA requirements with respect to door dimensions. To retain a degree of flexibility in the proposed solution space, the *door detection* node is parameterizable and door candidates are being kept only if they have a height larger than $\mathsf{h}_{\min}$ (the height of the robot) for example. The width of a candidate has to respect the minimally imposed ADA width for wheelchairs, while its maximum size can be selected based on a heuristic assumption. An implementation example would be to use a maximum width of 1.4m. In addition, each of the two door edges on the side of the door has to have a length of at least $\mathsf{h}_{\min}$. The task executive can change the value of any of these parameters online. Figure 11.5 presents two examples for door detection in cluttered environments using the previously described algorithm. Note that any of the above parameters could be automatically learned from a set of given training examples.

Fig. 11.5 Examples of door candidates identified successfully in different point cloud datasets

11.2 Detecting Handles

The *handle detection* node is invoked for restricted portions of 3D space, usually in the proximity where a door candidate has already been detected. To retain the same generality level present in the *door detection* node, the handle identification method operates and extracts the handle from the same point cloud data $\mathcal{P}$, without making use of additional data sources such as camera images. However, due to the nature of handles in general, such as their extremely thin geometrical structure and the materials they are made of, the sampled data representing them is extremely sparse and noisy. In addition, the laser rays that hit the metallic part of the handle sometimes sample points very poorly with geometry coordinates in the door or behind the door instead of in front of it. Taken together, these issues bring additional complexity to the handle identification problem, and lead to situations where it is impossible to perform a pure geometric segmentation.

To solve this, the proposed algorithms combine the sparse geometrical structure with the additional information provided by the intensity (or better said, surface reflectivity) data acquired and present in the laser scan. As shown in the following, this combination increases the robustness of the handle segmentation, and provides solutions which geometry alone would not be able to solve. Figure 11.6 presents both the intensity and geometry variations for a handle selected from a dataset $\mathcal{P}$. The main computational steps used by the *handle detection* node are presented in Algorithm 11.2.

The algorithm starts by segmenting a subset of points $\mathcal{P}_s$ from the entire point cloud $\mathcal{P}$, which could potentially contain a handle. The segmentation is performed using the information provided by the task executive, which states what doors have been detected and offers their geometrical parameters to the *handle detection* node. In particular, for a given door model $\mathsf{d} \in \mathcal{D}$ with $\mathcal{D}$ being the set of all detected doors for a dataset $\mathcal{P}$, the algorithm selects $\mathcal{P}_s = \{\boldsymbol{p}_i \mid \boldsymbol{p}_i \in \mathcal{P}, \boldsymbol{p}_i \subset \mathsf{V}\}$, where V represents the volume of a 3D polygon created from the bounding 2D polygon of d translated along the plane normal with $\pm h_{\mathrm{d}}$. The parameter h_{d} is given by the ADA requirements as the maximum distance from a door plane where a handle could be located. A simplification of the above is to get all points $\boldsymbol{p}_i$ whose distance from the

Algorithm 11.2. The main computational steps for the detection of door handles

$\mathsf{h}_{\mathrm{ADA}}, \mathsf{d}, \mathcal{P}$ *// ADA door handle requirements, door candidate, input point cloud*
for all $\boldsymbol{p}_i \in \mathcal{P}$
  $d_i = \mathrm{dist}(\boldsymbol{p}_i, \mathsf{d}), p_{r_i} = \mathrm{proj}(\boldsymbol{p}_i, \mathsf{d})$ *// distance and projection of* $\boldsymbol{p}_i$ *on* d
  if $(d_i \leq h_{\mathsf{d}}) \wedge (p_{r_i}$ inside $\mathsf{d})$ *// check if close to and inside door bounds*
    if $\mathsf{F}(\boldsymbol{p}_i, \mathsf{h}_{\mathrm{ADA}})$ *// does* $\boldsymbol{p}_i$ *respect the ADA constraints?*
      $\mathcal{P}_s \leftarrow \boldsymbol{p}_i$ *// add* $\boldsymbol{p}_i$ *to* $\mathcal{P}_s$
$(\mu_h, \sigma_h) \leftarrow \mathsf{F}(\mathcal{P}_s)$ *// estimate statistics for intensity distribution*
$(\mu_c, \sigma_c) \leftarrow \mathsf{F}(\mathcal{P}_s)$ *// estimate statistics for curvature distribution*
$\alpha_{\max} = 7$ *// maximum number of* $\pm$ *standard deviations to check*
for all $\alpha_i \in \{0 \cdots \alpha_{\max}\}$
  $N_{i_h} = \mathsf{F}(\mu_h \pm \alpha_i \cdot \sigma_h)$ *// get number of points* N_{i_h} *outside* $\mu_h \pm \alpha_i \cdot \sigma_h$
  $N_{i_c} = \mathsf{F}(\mu_c \pm \alpha_i \cdot \sigma_c)$ *// get number of points* N_{i_c} *outside* $\mu_c \pm \alpha_i \cdot \sigma_c$
$\mathsf{t}_h \leftarrow \mathsf{F}(N_{i_h})$ *// compute the trimean for number of points* N_{i_h} *distribution*
$\mathsf{t}_c \leftarrow \mathsf{F}(N_{i_c})$ *// compute the trimean for number of points* N_{i_c} *distribution*
$(\alpha_{b_h}, \mathcal{P}_h) \leftarrow \mathsf{F}(\mathsf{t}_h, N_{i_h})$ *// estimate the best cutting* α_{b_h} *value, and* $\mathcal{P}_h$
$(\alpha_{b_c}, \mathcal{P}_c) \leftarrow \mathsf{F}(\mathsf{t}_c, N_{i_c})$ *// estimate the best cutting* α_{b_c} *value, and* $\mathcal{P}_c$
$\mathcal{P}_f = \mathcal{P}_h \cap \mathcal{P}_c$

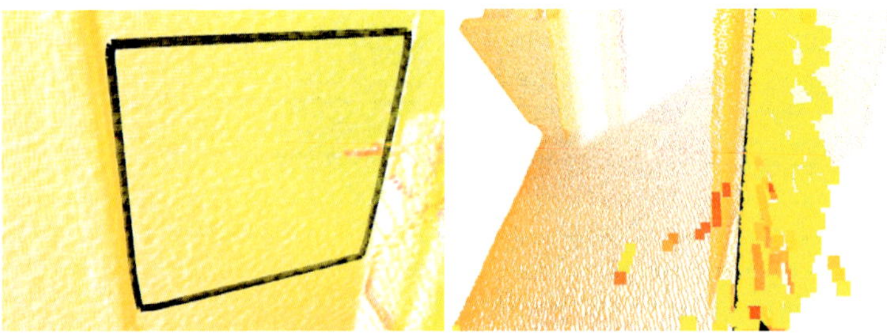

Fig. 11.6 Intensity and geometry variations for a selected handle. Left: view from the front, right: view from the top. Notice the extremely sparse geometric handle structure sampled in front of the door and the resultant points with intensities (remission information) belonging to the handle but sampled on the door plane, on the right part of the figure.

plane model of d is smaller than h_{d}, and check whether their projection on the plane falls inside the bounding polygon of d. Figure 11.7 presents the segmented $\mathcal{P}_s$ from a larger sequence, together with the ADA vertical bounds on where a door handle must be located.

Fig. 11.7 The selected $\mathcal{P}_s$ (represented with blue color) that potentially contains the door handle, based on a previously given door model, and the ADA door handle requirements

For each point $\boldsymbol{p}_i$ in $\mathcal{P}_s$, a neighborhood $\mathcal{P}_i^k$ around it is selected, and the surface curvature γ_p at $\boldsymbol{p}_i$ is estimated from the eigenvalues $\lambda_0 \leq \lambda_1 \leq \lambda_2$ of the covariance matrix corresponding to $\mathcal{P}_i^k$ as follows:

$$\gamma_p = \frac{\lambda_0}{\lambda_0 + \lambda_1 + \lambda_2} \tag{11.2}$$

Figure 11.8 presents the distribution of curvatures along the $\mathcal{P}_s$ set of points, with a large percentage of the points having a surface curvature close to 0 (*i.e.*, planar). The points having a spike in curvature space are potentially part of candidate handle clusters.

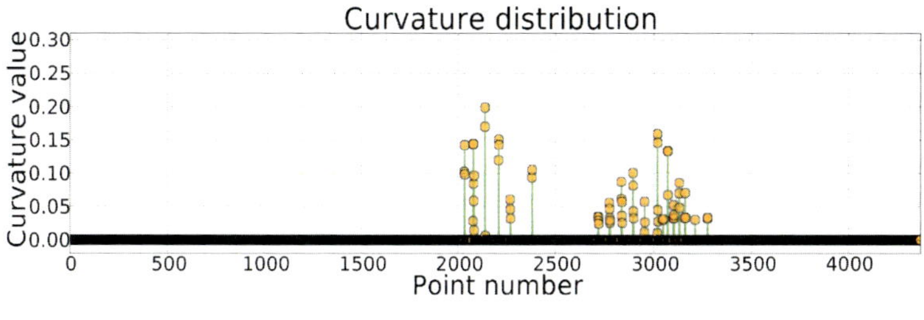

Fig. 11.8 The distribution of surface curvatures over a point cloud dataset. The majority of the points have a curvature value close to 0 which indicates that they lie on a planar surface.

Given the door handle's signature in intensity but also in curvature space, the system performs an analysis of the distribution of values over these spaces for all the points in $\mathcal{P}_s$. Because it is guaranteed that most of the points in $\mathcal{P}_s$ are situated on the door plane and thus part of the door itself, the resultant intensity distribution will have a highly peaked mean μ_h (see Figure 11.9).

Therefore, in principle all the points $\boldsymbol{p}_i \in \mathcal{P}_s$ whose intensity value is outside $\mu_h \pm \alpha_h \cdot \sigma_h$ could be selected, where σ_h represents the standard deviation of the aforementioned intensity distribution, and α_h is a user given parameter. Similarly, another set of points $\boldsymbol{p}_j$ could be selected for the curvature distribution, if the surface curvature value γ_{p_j} of $\boldsymbol{p}_j$ is outside $\mu_c \pm \alpha_c \cdot \sigma_c$, where μ_c and σ_c represent the mean and standard deviation of the curvature distribution, and α_c is a user given parameter respectively.

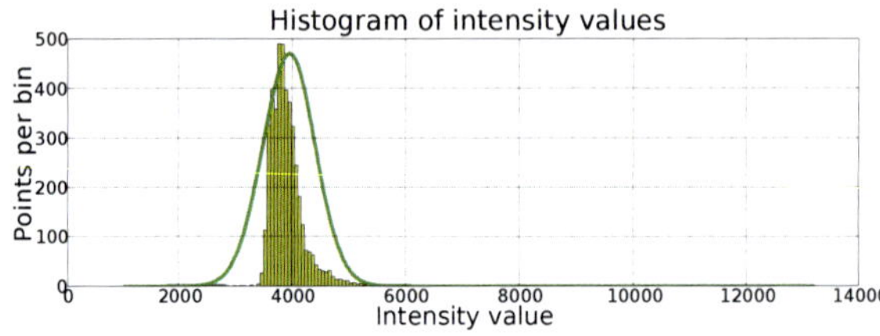

Fig. 11.9 The distribution of intensity values over a point cloud dataset. Notice the highly peaked mean μ.

Though the above formulation depends on two parameters only (α_h and α_c), it would be great to automatically determine their values for any door. To do this, the analysis makes use of Chebyshev's inequality which states that for any distribution that has a mean and a variance, at least $1 - 1/\alpha^2$ points are within α standard deviations from the mean. Selecting $\alpha_{\max} = 7$ accounts for at least 98% of the values. Thus the algorithm could iterate over the space of standard deviations with a given size step, and estimate the number of values falling outside of some parameterized interval. In detail, for each $\alpha_i \in \{0 \cdots \alpha_{\max}\}$, the number of points N_i from $\mathcal{P}_s$ falling outside of the interval $\mu \pm \alpha_i \cdot \sigma$ is computed, leading to the creation of two additional distributions (one for intensity values and one for curvature values) over the N_i space.

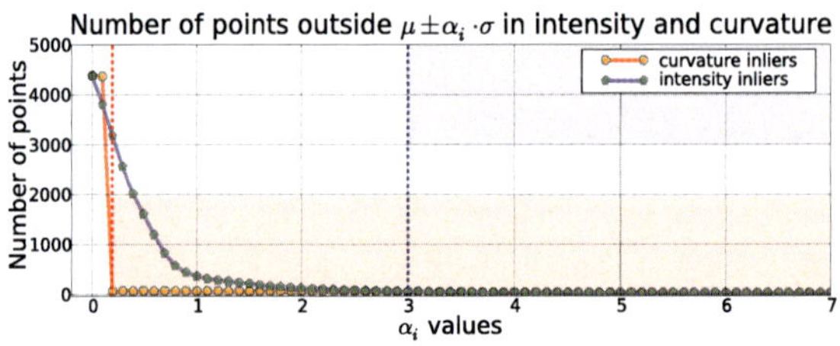

Fig. 11.10 The number of points falling outside the interval $\mu \pm \alpha_i \cdot \sigma$ for different α_i values, for the intensity and curvature distributions. The two estimated cutting points t_h and t_c are represented with dashed vertical lines.

Figure 11.10 presents the variations of the number of points N_i outside the interval $\mu \pm \alpha_i \cdot \sigma$ for different α_i values in both intensity and curvature spaces. The goal of this analysis is to find a minimal set of points which has extreme values in both spaces. As it can be seen from the two plots, the number of values drops really quickly as α_i grows, especially in the curvature space where almost 99% of the points are around the mean μ_c. To compute a good cutting value, the algorithm makes use of a robust L-estimator, namely the trimean $\mathtt{t}$ of each distribution, and selects the best $\alpha_{c_b}, \alpha_{h_b}$ values in both cases as the values closest to the trimeans. If $\mathcal{P}_c$ and $\mathcal{P}_h$ are the point sets outside the intervals $\mu_c \pm \alpha_{c_b} \cdot \sigma_c$ and $\mu_h \pm \alpha_{h_b} \cdot \sigma_h$ respectively, then the final set of point handle candidates $\mathcal{P}_f$ is obtained as:

$$\mathcal{P}_f = \mathcal{P}_c \cap \mathcal{P}_h \tag{11.3}$$

The actual handle is obtained by first projecting $\mathcal{P}_f$ on the door plane, and then fitting the best horizontal line (largest number of inliers) along the door axis in it using RMSAC. Figure 11.11 presents the resultant $\mathcal{P}_f$ handle inliers for the dataset in Figures 11.6 and 11.7.

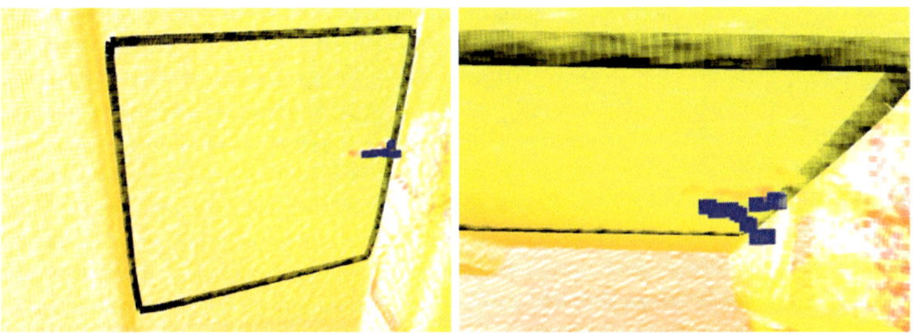

Fig. 11.11 Resultant handle inliers using dual distribution statistics analysis

11.3 System Evaluation

The door and handle detection algorithms compute the 6D pose (position and orientation) of the door and the door handle respectively, together with their geometric bounds. The results are then validated in a subsequent step where the robot approaches the door and attempts to grasp the door handle. To assess the performance of the proposed perception methods, the grasping is realized in "open loop", with no additional sensory feedback used, in two phases. In the first phase the robot base navigates to a pose in front of the door, from where the door handle is within the workspace of the robot arm. The base navigation node moves the base towards the desired pose, based on measurements from the wheel odometry and an Inertial Measurement Unit (IMU). In the second phase, the robot arm moves along a collision free trajectory to position the robot gripper on the door handle. The success rates for the door and handle detection is then measured based on the robot's ability to achieve a caging grasp around the door handle. Figure 11.12 depicts an example of a door and handle candidate identified successfully in a point cloud dataset.

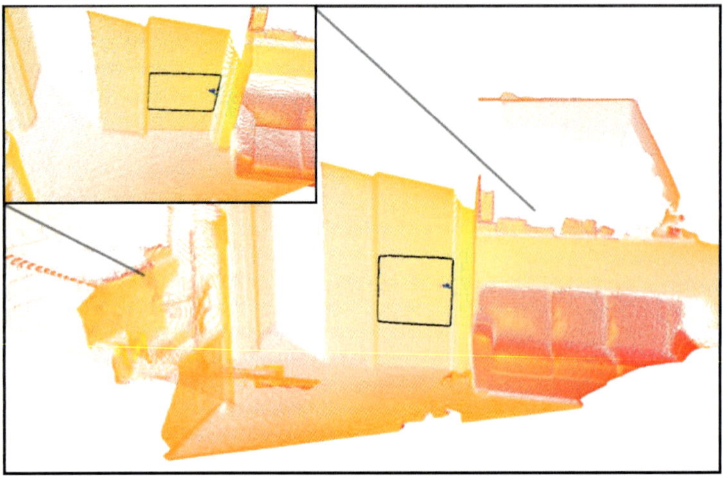

Fig. 11.12 Example of a handle candidate identified successfully in a point cloud dataset

The presented validation method allows for small misalignments of the gripper with respect to the door handle, while still achieving a successful grasp. On the PR2 robot for example, with the gripper fully opened, the distance between the fingertips is approximately 8cm. For a door handle with a thickness of 2cm, this results in a maximum allowed vertical gripper misalignment of 3cm. The allowed left-right gripper misalignment is bounded by the width of the door handle lever to a maximum of 3.5cm. The misalignments observed in real world experiments are typically within 2cm.

To validate the proposed framework, the system was evaluated on more than 50 different situations, where doors were scanned in different states (open/closed/half open) and from different angles. Additionally, Willow Garage's Milestone 2[1] proved to be a tough testbed for the proposed algorithms, with the PR2 robot having to identify, navigate, and open various different doors in its task of finding several power plugs where it could recharge its batteries in less than 1 hour. The point cloud datasets of the experiments and a demonstration video explaining the acquisition and processing of point cloud data for the purpose of door and handle identification and validation can be downloaded from the Willow Garage web site.

The datasets have been acquired in varying light conditions, sometimes completely in the dark, thus accounting for a very large variation of situations where 2D image-based segmentation would fail. Both the door and handle detection performed extremely well (averaging over 95%) and succeeded in segmenting and identifying the object components given valid input data which respected the system constraints. Tables 11.1 and 11.2 present examples of the estimation results obtained on subsets of the above mentioned datasets. The first 4 rows of both tables present valid results, verified using the methods presented above.

[1] `http://www.willowgarage.com/blog/2009/06/03/watch-milestone-2`

Some of the situations which could not be entirely handled by the proposed methods include examples such as the ones shown in the bottom row of the tables. For example, in Table 11.1, the following examples (from left to right) are shown: i) a wall which respects the geometric constraints of a door being segmented as a door candidate; ii) only one out of two doors being successfully identified in the data, iii) and iv) unsuccessful door segmentation due to no geometric differences between the walls of the room and the door itself. The latter is an issue of the door segmentation approach, but unfortunately it cannot be solved by a revised version of the algorithm, simply because the geometry and intensity levels of the door are indistinguishable from the wall. One possible solution would be to include stereo data available on the PR2 robot and attempt to detect the door edges in RGB space.

A few failure cases for the handle segmentation method are also presented in the bottom row of Table 11.2. In the first example, the door candidate given by the executive was a refrigerator door, but due to variations in geometry on the door handle, only a partial subset of it was detected and returned. The next two cases are not

Fig. 11.13 Handle identification failure due to an extremely low (3) number of point hits on the actual door handle

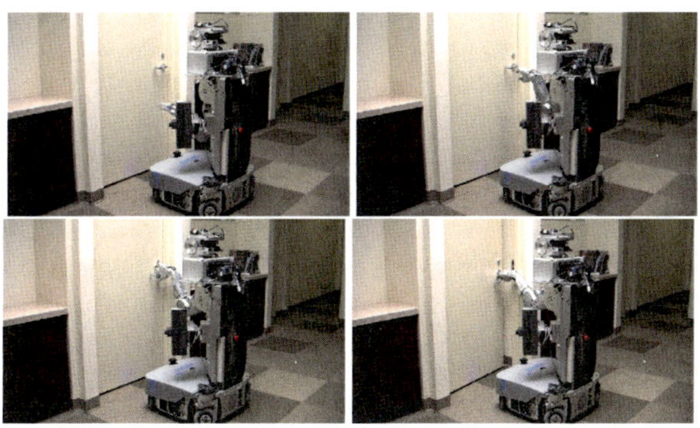

Fig. 11.14 Snapshots taken during one of the experimental trials concerning door identification and opening

Table 11.1 A subset of 20 datasets used to test the door detection algorithm. The first 4 rows show successful identification cases, while the last row presents difficult segmentation situations where: a) a part of the wall resembling a door has been selected as a candidate; b) only one of out two doors have been detected; c) and d) the door is not detected due to under-segmentation.

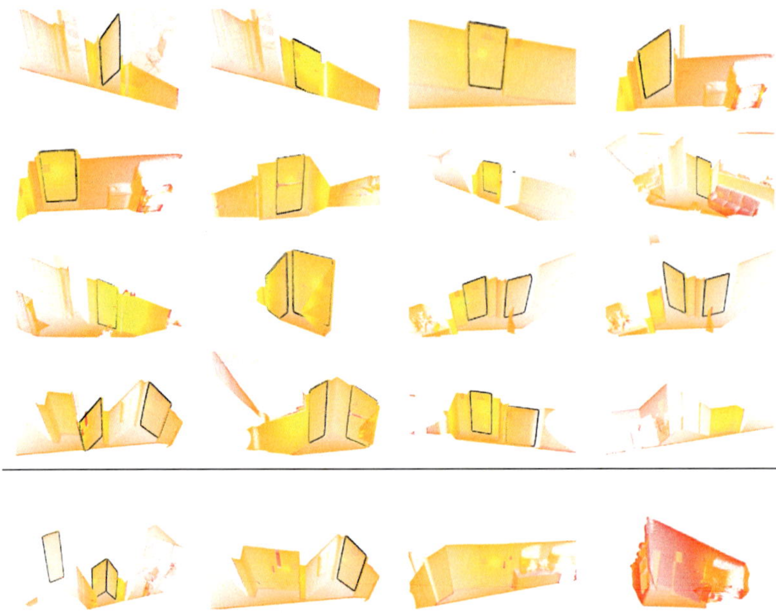

a b c d

really failures, as the candidates given are in fact walls, and thus the handle detection algorithm returns an empty list of handles. The last example however presents an under-segmentation of the points belonging to the door handle. The explanation of this error is given in Figure 11.13, where a closeup of the same dataset is shown. Due to the fact that the sampled geometry of the handle contains only 3 point hits (as presented in the right part of the figure), the handle extraction algorithm rejects the candidate. An immediate solution to this problem is to take a new scan and simply concatenate the previous dataset with the new one to obtain more point hits on the door handle.

As shown in the results presented, the proposed methods are not influenced by the door opening angle or the handle type, as long as a basic constraint is respected: there exists a distinguishable difference in intensity and curvature between the actual handle and the door plane. Figure 11.14 presents a few example snapshots taken during one of the experiments concerning door identification and opening.

Table 11.2 A subset of 20 datasets used to test the handle detection algorithm. The first 4 rows show successful identification cases, while the last row presents challenging situations for: a) a refrigerator door; b) and c) testing the handle detection on two walls which resemble door candidates; d) unsegmented handle for a regular door.

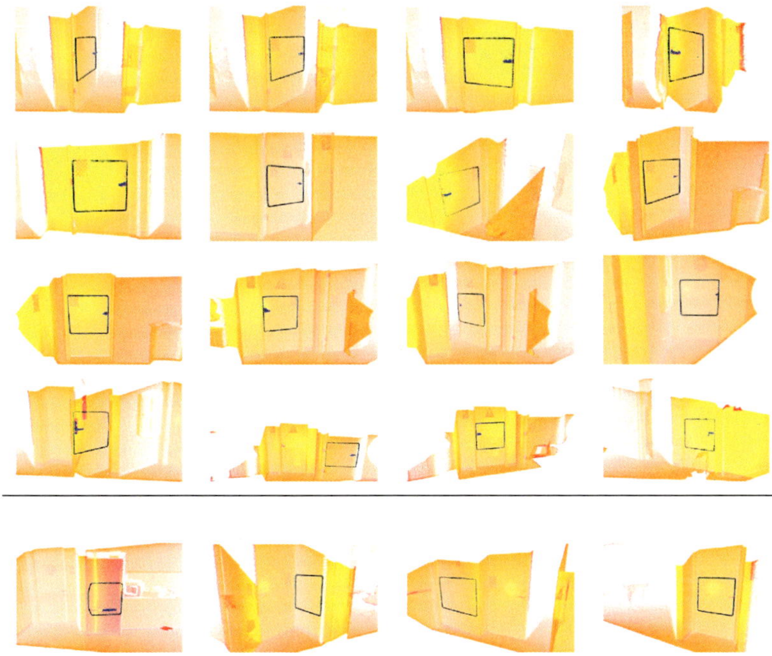

a b c d

11.4 Summary

This chapter presented a set of robust methods for the problem of door and handle identification from noisy scanned 3D data in indoor environments. By refraining from using camera images and constraining the search space using the ADA (American Disability Act) imposed requirements, the proposed system can successfully identify doors and handles in situations where methods based on 2D images would fail, such as varying light conditions or no light sources at all. By making use of the proposed dual intensity-curvature distribution analysis, the segmentation parameters are automatically identified from the sensed data, without the need to training a machine learning classifier.

Chapter 12
Real-Time Semantic Maps from Stereo

> *"It's hardware that makes a machine fast. It's software that makes a fast machine slow."*
>
> CRAIG BRUCE

Creating rich, meaningful map representations of the world from stereo data poses two interesting questions: i) given that stereo depth data is traditionally not as precise as laser measurements, what is the level of detail that a stereo-based semantic map could have; and ii) how can a mapping system able to process the large quantities of data that stereo cameras produce (*e.g.* frame rates of 30 fps) in an efficient real-time manner should be designed.

A quick attempt at building system architectures such as the ones presented in Chapter 3 would result in catastrophic failures. Because of its different working principles, depth as acquired by stereo triangulation is highly sensitive to the point correspondences found in the two sets of 2D images, which are sensitive to lighting conditions and texture information themselves. Too many mismatches will lead to the resultant point clouds having too many outliers, which makes their registration extremely difficult. Additionally, stereo captured data might not always approximate the real object shape required for 3D geometric reconstruction well enough. To better illustrate the above, Figure 12.1 presents the resultant point clouds from a textured cylindrical shape acquired using a SICK LMS400 laser (left) and a Videre stereo camera (right), together with their surface normal estimates. The obvious that this analysis wants to capture is that the estimated surface normals from the laser acquired dataset do tend to capture the cylinder's geometry, while the ones from the stereo are more chaotic and thus do not model the underlying surface geometry with the same fidelity. So in some sense, stereo does not provide the same guarantees on the resulting output depth data. This is an important limiting factor for the application of 3D geometric model fitting in stereo point cloud data.

Though in all fairness, better stereo matching techniques exist (see [NSG08] for a recent review), they are not yet suitable for online real-time applications. It is therefore imperative to find a system architecture that can acquire point clouds from a stereo camera and build a 3D map annotated with semantic labels in a

R.B. Rusu: *Semantic 3D Object Maps for Everyday Robot Manipulation*, STAR 85, pp. 177–196.
DOI: 10.1007/978-3-642-35479-3_12

very fast way, thus enabling the development of reactive applications with constrained time requirements. To bound the problem, the hardware perception system is restricted to the use of a stereo camera setup with a minimal number of extra devices.

Fig. 12.1 Example of surface normal estimates (blue) for a textured scanned cylinder using phase-shift laser (left) and a stereo camera (middle). The right part of the figure presents the environment used with the cylindrical pipe marked.

A proposed system architecture example is shown in Figure 12.2. In contrast to the previously presented architectures from Chapter 3, the integration of individual stereo-acquired point cloud views acquired into a consistent global model is performed using globally estimated poses from a Visual Odometry module. Keeping track of the camera pose at all times relies on a combination of feature tracking in subsequent 2D images and local poses obtained from an IMU attached to the system. By using these precise poses over time, the registration of individual point clouds becomes rather trivial, and as previously explained in Chapter 5, most often it suffices to simply apply the rigid transformation between two camera poses v_{p_1} and v_{p_2} to a point cloud acquired from v_{p_2} in order to align it with its sibling acquired from v_{p_1}.

Due to the previously mentioned inability to model non-linear geometric surfaces accurately enough from stereo data, the proposed processing pipeline relies on local approximation of the world using convex planar patches. The core 3D processing module would therefore take registered point cloud models as input, and apply a series of geometric processing steps to them, resulting in locally refined polygonal models. Based on heuristic semantic rules either given or learned, these models can then be annotated with a certain surface or object class, useful for the application at hand.

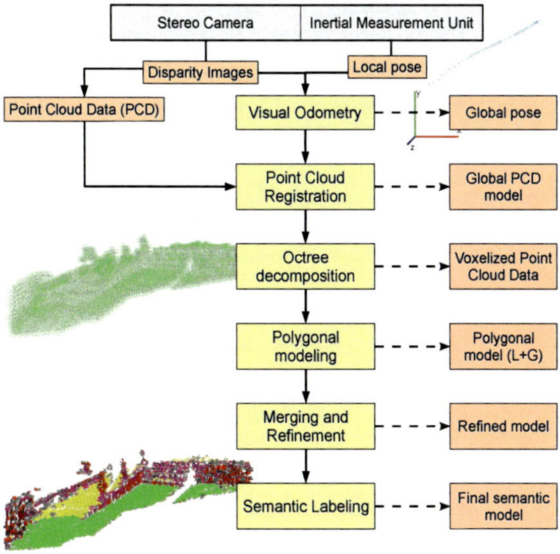

Fig. 12.2 An architecture for real-time mapping using stereo data. The algorithmic processing steps or modules are shown in light yellow boxes, while the resultant data structures are shown in light red.

12.1 Leaving Flatland Mapping Architecture

A variant of the above proposed architecture was used in the Leaving Flatland project [RSM+09, MRS+09, RSM+08], for the purpose of real-time mapping and planning where to navigate with a RHex (Robotic Hexapod) robot [SBK01], using stereo data. The presented mapping system is therefore only a subcomponent of a larger, integrated architecture, including motion planning and navigation routines (see Figure 12.3). Figure 12.4 depicts the rugged RHex mobile robot from Boston Dynamics, a highly capable six-legged mobile robot, able to traverse various types of challenging environments at high speed.

The main goal of the Leaving Flatland project is to surmount the challenges of closing the loop between autonomous perception and action on challenging terrain. This includes extremely efficient mapping, motion planning, and navigation methods that would allow mobile robots to travel in the same environments, at the same speed, and with the same reliability and flexibility as humans do without requiring any prior preparation. Such robots are necessary for many tasks that require long-term navigation such as environment exploration, rescue operations, performance of scientific experiments, and object delivery in an unknown environment. As previously mentioned, the target robotic platform used to validate the theoretical approach is a RHex mobile robot. RHex can run at high speed on flat ground, crawl over rocky terrain, and climb stairs. Equipped with a stereo camera and an

IMU, the robot can keep track of its position and orientation using Visual Odometry techniques. The stereo depth data is also used to build a 3D model of the environment that is continuously updated as the robot moves. Periodic model updates are sent to a multi-region motion planner, which generates 3D trajectories that are to be executed by the robot. These trajectories consist of a sequence of primitive gait commands which are sent to the robot's on-board controller. The planner is typically asked to reach a nearby goal within the robot's visual range, usually chosen by either a human operator or some supervisory behavior. Long-range goals are achieved by frequent planning. In order to increase the mobility of the robot and help the planner select the most appropriate gait for the terrain the robot is to navigate, the 3D models are annotated with semantic labels. The motion planner uses the inferred terrain labels to decompose the terrain locally into 2D regions and selects fast planning techniques within each region.

The evaluation of all the above components on a variety of real world datasets showed an extremely favorable computational performance for high-speed autonomous navigation, with real-time mapping updates averaging about 250ms. Note that it is not necessary to update the map unless there is significant change in the view of the camera. The motion planner can be queried approximately once or twice a second depending on the environment's complexity.

The objective of the 3D mapping module is to build and update a model of the scene in a fixed global coordinate frame, which can then be used for planning a 3D path. The global coordinate frame can be chosen arbitrarily and is fixed for the duration of each sequence. The mapping module uses the image-disparity data obtained from the stereo sensor as input. A Visual Odometry (VO) module is used in order to keep track of the pose of the robot. The pose with respect to the fixed global coordinate frame can be used to register the point cloud data to a fixed coordinate frame. The registered data is passed through a mapping pipeline that builds and updates a 3D model of the scene. To explain the role of each sub-module, the following notations are introduced:

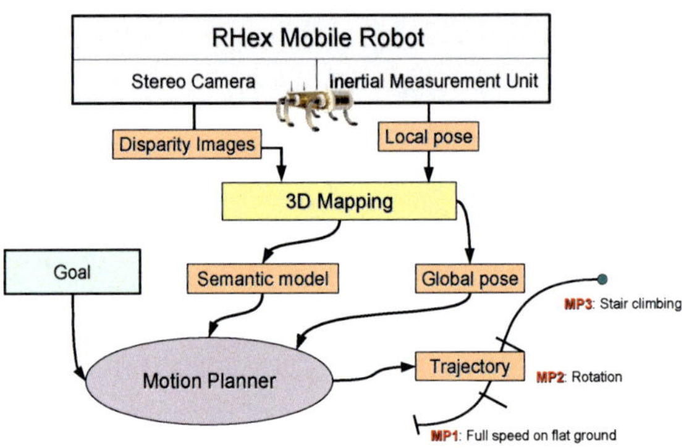

Fig. 12.3 The general architecture of the Leaving Flatland project [RSM+09]

- $\mathcal{P}^{(t)} = \{\boldsymbol{x}_1^{(t)}, \boldsymbol{x}_2^{(t)}, \ldots\}$ is the registered point cloud data set, *i.e.*, the set of 3D points in the global coordinate frame acquired at time t. $\boldsymbol{x}_i$ is a 3D point having coordinates $(\, x_i \; y_i \; z_i \,)'$ in the global coordinate frame, with an optional color vector $(\, r_i \; g_i \; b_i \,)'$.
- $\mathcal{M}_L^{(t)}$, $\mathcal{M}_G^{(t)}$, and $\mathcal{M}_F^{(t)}$ denote the local, global, and final polygonal models at time t.

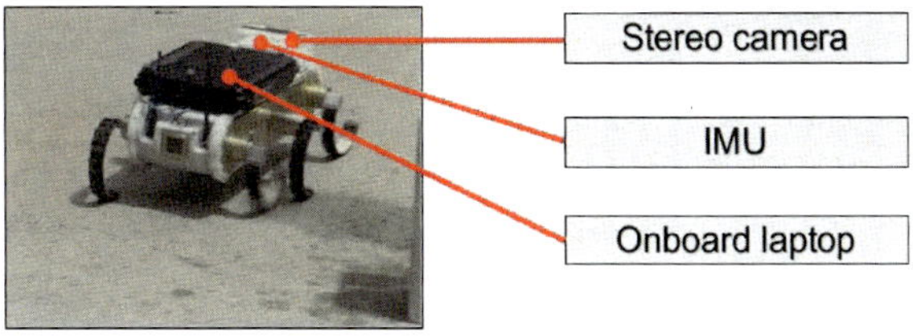

Fig. 12.4 The RHex (Robotic Hexapod) mobile robot used in the Leaving Flatland project

At each time instant t, the 3D mapping system performs the following steps:

1. Decompose the point cloud data $\mathcal{P}^{(t)}$ using a dynamic octree (see Section 12.1.2).
2. Compute the local model $\mathcal{M}_L^{(t)}$ from decomposed $\mathcal{P}^{(t)}$ (see Section 12.1.3).
3. Merge the local model $\mathcal{M}_L^{(t)}$ with the global model from previous frame $\mathcal{M}_G^{(t-1)}$ and compute the updated and refined global model $\mathcal{M}_G^{(t)}$ (Section 12.1.4).
4. Add semantic labels to compute the final model $\mathcal{M}_F^{(t)}$ (Section 12.1.5).

The output of the mapping system is represented by a 3D polygonal model annotated with semantic labels. A detailed description of the mapping components listed above is described in the following sections. For a more general discussion, please refer to Chapter 5.

12.1.1 Visual Odometer

The accuracy of the computed global map depends directly on the precision with which the pose of the robot is evaluated with respect to its previous location in the world. This evaluation needs to be performed online as robustly as possible while the robot is moving. Another way to formulate the problem is to determine the 3D robot pose increment (*i.e.*, the 3D transformation) with respect to its former location, such that the overlap between the current and former camera views has a minimal error in an Euclidean distance metric sense. This process is also called *registration* and its accuracy determines the quality of the final model. Instead of employing a classical registration framework (*e.g.* Iterative Closest Point), the solution adopted for Leaving Flatland uses a Visual Odometer (VO).

As input, the VO takes a series of monocular and disparity images and computes the camera motion (3D rotation and 3D translation) between those views. This motion is estimated by extracting distinctive features from each new frame in both the left and the right camera images and matching them against each other, and also against the features extracted in the previous frames. As previously presented in Chapter 5, a Visual Odometry system can work reasonably with different types of 2D features. An important consideration however, is to use features that are stable across viewpoint changes. While SIFT [Low04a] and SURF [HBG06] are the feature detectors of choice, they are not suitable for real-time implementations (15 Hz or greater). Instead, Leaving Flatland uses a novel multiscale center-surround feature detector called CenSurE [AKB08], due to its requisite stability and extremely efficient computational properties.

From these uncertain feature matches obtained, a consensus pose estimate is recovered using a RANSAC method [FB81]. Several thousand relative pose hypotheses are generated by randomly selecting three matched non-collinear features, and then scored using pixel reprojection errors. If the motion estimate is small and the percentage of inliers is large enough, the frame is discarded, since composing such small motions increases error. This pose estimate is then refined further in a sparse bundle adjustment (SBA) framework [ESN06].

Given the pose of the robot in an arbitrary global coordinate system, each point cloud $\mathcal{P}_t$ computed in the robot's camera frame can be registered to the fixed global coordinate system. In detail, using the pose of the camera at frame t, the transformation (R_t, t_t) from the camera coordinate frame where the point cloud data resides to the global coordinate frame can be computed. This transformation is then applied to the point cloud data set acquired at t to obtain $\mathcal{P}^{(t)}$. Figure 12.5 illustrates the Visual Odometry registration process for two individual local point clouds obtained from different view points.

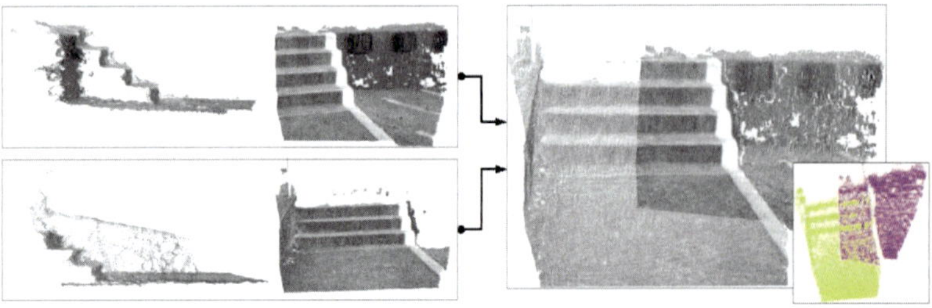

Fig. 12.5 Point cloud registration using Visual Odometry: The two images on the left represent data from two frames approximately 2 seconds apart. Both frames have been registered to a global coordinate frame using the Visual Odometer module and the result of the registration is illustrated in the images on the right.

Because the VO algorithm computes new camera poses at about 10 Hz, the variation between two consecutive poses is typically small, and therefore it would be highly inefficient to update the 3D model with the same frequency.

(a) *stairs* sequence (b) *dumpster* sequence

Fig. 12.6 Globally aligned point cloud data for two datasets (top image is birds-eye view and the bottom image is the close-up view). (a) *stairs* ($\approx$ 2000 frames), (b) *dumpster* ($\approx$ 3000 frames).

Therefore, a convention is made to update the model only if the change in the camera pose since the last update is greater than a threshold (an implementation example would be 0.1m for the translation and 0.1rad for the rotation). Figure 12.6 shows results obtained on two different datasets (*dumpster* and *stairs*) after the alignment of approximately 3000 and 2000 stereo frames respectively.

The accuracy of the global map depends on the quality of the Visual Odometry pose estimate. Due to excessive camera motions, and in particular to rapid rotational accelerations of the robot, some images can appear as blurred (see Figure 12.7). In addition, some areas of the environment do not contain enough features for the VO tracker. If the system skips too many frames, the global pose estimate will be erroneous. If several failures from VO occur consecutively, the system could lose track of the robot motion and the integrity of the 3D model could be compromised.

These issues can be partly dealt with by integrating an Inertial Measurement Unit (IMU) sensor in the Visual Odometry system. The IMU can provide local pose information which can be used to correct and keep the VO pose estimates consistent. Figure 12.8 presents the corrections in roll (bottom left) and pitch (bottom right) before and after the IMU integration in the VO module. Even if some error in pitch persists, the drift on z is reduced by approximatively 65% (top right). In the sequence in Figure 12.7, the robot navigates a ramp and the initial and final positions of the robot are on level ground. The initial and final values of the Z-coordinate, pitch and roll should therefore be zero.

Fig. 12.7 Example of challenging frames for the Visual Odometry system

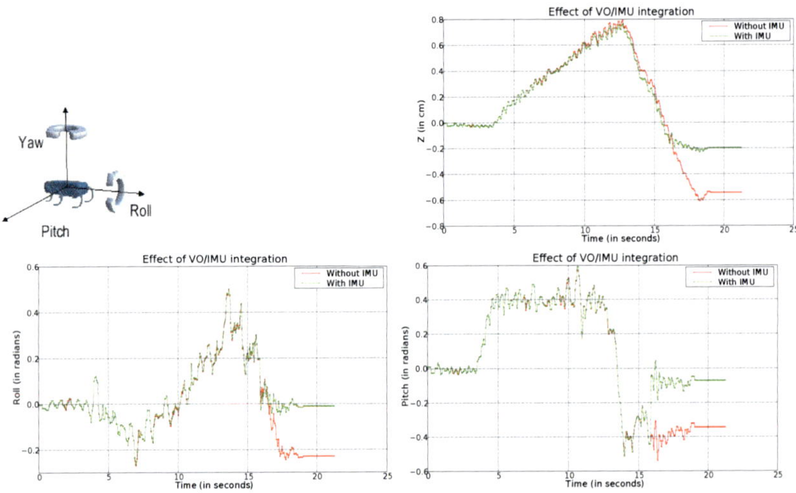

Fig. 12.8 Global pose corrections by integrating an IMU into the Visual Odometry system

12.1.2 Spatial Decomposition

The point cloud dataset obtained from the stereo sensor at each frame, $\mathcal{P}^{(t)}$, contains hundreds of thousands of points. As the robot moves, it acquires points from regions that overlap the existing model as well as from previously unexplored regions. From a real-time processing point of view, the system faces two challenges. First, it needs to efficiently handle and represent redundant data, as storing or processing the accumulated set of points becomes unmanageable after only a few data frames. Second, it needs to be able to process and update only the relevant parts of the model. The second aspect becomes important as the size of the model grows and it becomes impractical to update the model as a whole. The choice of data representation is therefore critical to the efficiency of the entire system. Point-based representations are generally unsuitable for other 3D data consumers such as motion planners, as many fast geometric queries need to be performed on the model, like distance and collision checks, etc.

A viable solution is to reconstruct the original surface which the sampled point cloud data represents as a set of polygons and use these as data representatives. The first step towards this surface reconstruction is to spatially decompose the data into chunks so that they could be processed faster. Therefore, at any given time, the existing model or data is referred to as *global*. The new data obtained at each time frame t is referred to as the *local* data and the corresponding model as the *local* model. The global polygonal model is updated whenever a new (local) data is obtained, at a frequency compatible with the speed of robot motion. As mentioned earlier, the system needs to have the ability to process or update only the relevant part of the global model, *i.e.*, the region that overlaps the local data.

Though different spatial decomposition techniques are possible (see Chapter 2.2), an octree structure is extremely suitable for the task at hand, because it natively supports different levels of details using its parent-children structure. The volume of interest in the data is therefore decomposed into several cells. The size of a cell can be set based on a number of factors determined by the application such as the required accuracy of the model, the expected terrain, etc, with implementation examples ranging from a few centimeters to tens of centimeters. To facilitate the merging of local and global models, an octree decomposition is performed so that the cells are perfectly aligned by snapping them to a common grid. This ensures that there exists a one-to-one correspondence between the cells in the local model ($\mathcal{M}_L^{(t)}$) and the cells of the global model ($\mathcal{M}_G$) wherever there is an overlap.

A local octree structure is obtained by performing a spatial decomposition of the local point cloud dataset in each frame ($\mathcal{P}^{(t)}$). The bounds of the local volume of interest are determined by the bounds of $\mathcal{P}^{(t)}$, as illustrated in Figure 12.9. The size of the cell at the leaf of the octree is fixed (l_{SIZE}) and the points that belong to each cell are represented by a suitable model as described in Section 12.1.3. The model for each cell is computed using points from only that cell and since the cells are non-overlapping, it is possible to perform this operation in parallel if desired.

(a) (b)

Fig. 12.9 Spatial decomposition using fixed-size octrees for two local point cloud datasets. Only the occupied cells are drawn.

For the first acquired frame, the bounds of the global octree will be the same as the bounds of the local octree. Subsequently, the global octree is grown dynamically to accommodate the point data from later frames.

The data points obtained are typically contiguous and occupy only a fraction of the volume of interest (which is a cube), thus making the octree representation an efficient way to both represent and access the data.

12.1.3 Polygonal Modeling

The main goal of the modeling step is to represent the large number of points in a manner that is both easy for the motion planner to use and also efficient in terms of storage and speed. The point cloud data is therefore converted into a local polygonal model thus reducing the size of the model by several orders of magnitude – from several hundreds of thousands of points to a few hundred polygons. As previously described in Section 12.1.2, each point is assigned to a specific cell in the local octree. For each occupied cell in the local octree, a set of convex polygons that represents the points contained within is estimated, thus approximating the entire world with a set of local planar patches. This represents the local model $\mathcal{M}_L^{(t)}$.

The decision of selecting planar patches decomposition over other surface reconstruction techniques is bound to the requirements of being able to update the surface representation as fast as possible. An ideal representation would obviously model the 2D surface concavity in an octree cell, but concave polygonal modeling through alpha shapes for example is still poised towards the choice of the alpha α parameter, and solutions for its automatic adaption for every type of surface is still a topic of research. These justify the use of planar convex polygons, which are good local approximations to the underlying regular geometric structures present in the scene. Moreover they are faster to compute over triangulation methods or concave hulls (alpha shapes), are geometrically stable, and do not require any thresholds or parameters.

The individual steps which lead to the creation of the local polygonal model $\mathcal{M}_L^{(t)}$ are listed in Algorithm 12.1. Note that since steps $5-10$ are computed independently for each cell, a parallel scheme that boosts the overall computation efficiency of the model by using multiple CPU cores [SMRR07] where available can be easily implemented. For most experiments, the values of the thresholds N_{min} and N_{poly} was set to $N_{\text{min}} = 5$ and $N_{\text{poly}} = 1$.

The search for planar models is performed using a RMSAC (Randomized M-Estimator SAmple Consensus) estimation method, which adds a $T_{d,d}$ test as proposed in [CM02] to a MSAC estimator [TZ00]. For each plane, the inliers are projected to the plane, and then a convex polygon which contains the projections is computed. The support of each polygon (*i.e.*, the number of inliers, area, etc) is stored for further refining operations.

Figure 12.10 presents the polygonal fitting results for a given scene in the *stairs* sequence, while Figure 12.11 presents the resulting polygonal models for the two datasets shown in Figure 12.6.

Algorithm 12.1. Computing local polygonal models for a given point cloud view

Require: a point cloud dataset $\mathcal{P}^{(t)}$
set N_{poly}, the number of polygons per cell
set N_{min}, the minimum number of points per cell
build a local octree representation O_L for $\mathcal{P}^{(t)}$
for every occupied cell i in O_L
  let $\mathcal{C}_i$ be set of all points in cell i
  if $|\mathcal{C}_i| > N_{\text{min}}$
    search for a maximum of N_{poly} planes in $\mathcal{C}_i$
    for each plane found j = 1:N_{poly}
      project all inliers $\mathcal{C}_i^j$ onto the jth plane
      construct a 2D polygonal hull which approximates $\mathcal{C}_i^j$ and add it to $\mathcal{M}_L^{(t)}$

Fig. 12.10 Point Cloud model simplification using polygonal fitting. From left to right: overlapping aligned point cloud views, combination point cloud data model and polygonal model, and the resulting polygonal model.

12.1.4 Merging and Refinement

The local model $\mathcal{M}_L^{(t)}$ is merged with the global model $\mathcal{M}_G^{(t)}$ at the local cell level. Depending on the chosen local octree cell size, one polygon per cell ($N_{\text{poly}} = 1$) might be sufficient. For each pair of overlapping octree cells from the local and global models, a comparison is made between the respective polygons based on their relative orientation and distance. The "distance" between polygons is determined by computing the integral of the distance of a point along the circumference of the smaller polygon from the plane of the bigger polygon. The integral is normalized by the circumference of the smaller polygon to obtain the distance of the smaller polygon from the bigger polygon. The orientation between the polygons is represented by the angle between their normals. Two polygons are judged to be similar if both the distance and the angle between them are less than some preset threshold values. From an implementation point of view, the distance threshold could be set to something like 2cm, together with an angle threshold $\approx 15^\circ$.

If two polygons are found to be similar, they will be replaced by a newer polygon, constructed such that it averages the two, based on an inlier re-weighting scheme. In detail, the best plane fit is computed using the vertices of both polygons weighted by the number of inliers in each. The points are then projected to this new plane

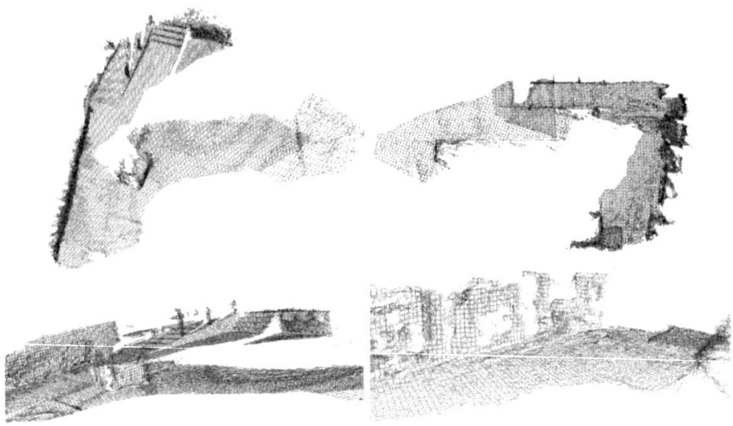

Fig. 12.11 Resultant polygonal models for two datasets: *stairs* - left ($\approx$ 2000 frames), *dumpster* - right ($\approx$ 3000 frames). Top: birds-eye view, bottom: close-up view.

and a convex polygon is fitted to the projections. If the polygons are dissimilar the model is replaced with the polygon that has the larger support. In the general case where $N_{\text{poly}} > 1$, the polygons are merged such that the number of polygons in the cell is capped at N_{poly}.

In order to perform the refinements, the algorithm considers only the cells in the global model that have been updated as a result of merging the polygons from the local polygon. Isolated polygons (*i.e.*, polygons that have no neighbors) are further removed through an outlier analysis. Because a single location can be observed several times during the mapping process, some surfaces may generate duplicate polygons in adjacent cells (see Figure 12.12). These duplicates are identified by searching for similar polygons in neighboring cells that are very close to each other.

The premise for reducing the polygonal model is based on the geometric relationships between adjacent polygons, namely the angle and the "distance" between them. If neighboring polygons are similar, the polygon with smaller support is removed. If, within a group of $2\times2\times2$ cluster of 8 cells, all the polygons are similar to each other, they are replaced with a single "super" polygon in order to simplify the representation.

The algorithm is in principle similar to the one presented in [WGS03]. However, the main difference lies in the precision of the polygonal simplification, which is in fact a weighted re-estimation of the original polygons, and thus more precise. Additionally, the algorithm performs lone polygon filtering in the same step. Experimentally, this procedure reduced the global number of polygons in $\mathcal{M}_G$ by a factor of 2.5 on average (see Figure 12.12).

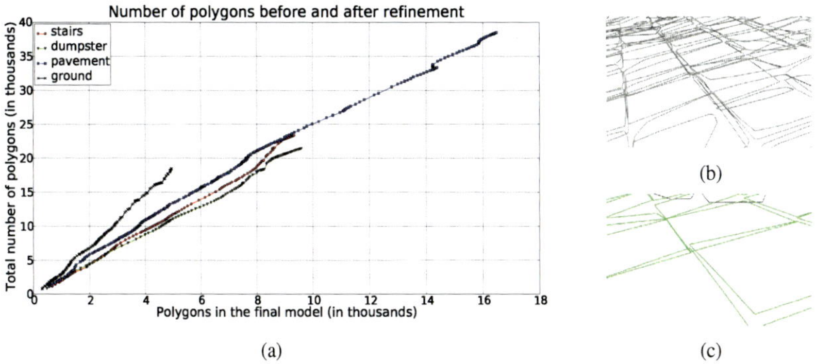

Fig. 12.12 Reducing the model complexity through merging and duplicate removal. (a) The polygonal reduction factor for 4 different datasets. (b) Model before refinement (c) Model after refinement.

12.1.5 Semantic Labeling

Since the proposed motion planner can manipulate several types of motion primitives, each of them adapted to a specific terrain type, it is important to help the planner to select the most appropriate primitive based on the type of terrain the robot has to navigate. To enable this, the polygonal model $\mathcal{M}_G^{(t)}$ is annotated with semantic class labels based on heuristic rules, thus creating the final model $\mathcal{M}_F^{(t)}$.

The labeling is performed by an analysis of the local 3D terrain geometry at and around each polygon. Though these rules are set heuristically, learning a classifier that automatically generates them is trivial, as their proposed feature space is easily separable. More sophisticated terrain types might be addressed by the methods of [MTJ+07, SWBR04, NSH03, RMB+08a], which account for more global geometric relationships. Leaving Flatland employs the following five object classes:

- **Level:** Polygons that form an angle less than 15° with the horizontal plane;
- **Ground:** Polygons that form an angle less than 15° with the ground plane and are connected to the ground plane;
- **Vertical:** Polygons that form an angle greater than 75° with the horizontal plane;
- **Stairs:** Groups of adjacent polygons that form steps, *i.e.*, belong to planes that are parallel or perpendicular to the horizontal plane;
- **Unknown:** Everything else.

To help RHex traverse small slopes, the set of *ground* polygons is expanded to include neighboring surfaces that respect the same properties within some bounds. In detail, once the set of *ground* polygons is labeled, a wave is propagated from them, such that at each iteration the neighbors of the seed *ground* polygons are checked to see if their geometry is similar to that of the ground plane. An implementation example would be to check if their angle is less than 15° from the ground plane, respectively if their distance is within 2cm from the seed polygon. All adjacent polygons respecting these constraints are also labeled as *ground*. With this ground propagation, a surface with a smooth inclination (up or down) is still considered flat ground. RHex can traverse small slopes and unevenness with ease, so ignoring minor irregularities helps speed up the motion planner. Figure 12.13 presents the semantic annotations for the scene in Figure 12.10, while Figure 12.14 presents the overall semantic models for the two datasets shown in Figure 12.6.

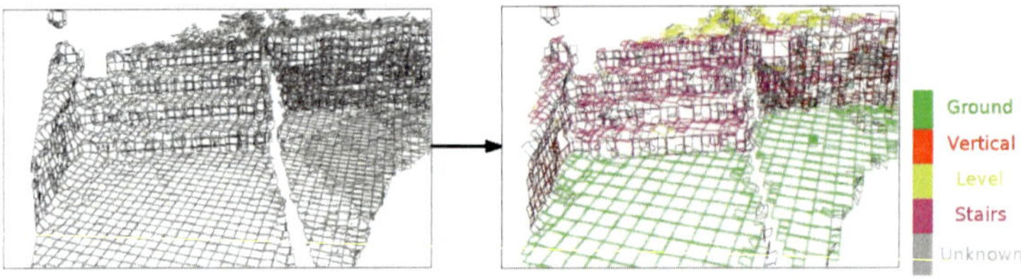

Fig. 12.13 Semantic labeling of polygonal surfaces based on heuristic geometrical rules. The labeled classes are: flat and level terrain (green and yellow), vertical structures (red), stairs (purple), and unclassified or unknown (gray).

Algorithm 12.2. An algorithm for labeling stairs from polygonal data

Require: a set of horizontal (m_1) and vertical (m_2) polygonal models

if size of m_1 and size of $m_2 \geq 1$
  compute the geometric centroids of m_1 and m_2 as c_{m_1} and c_{m_2}
  for each horizontal centroid in c_{m_1}
    search for nearest vertical neighbors in c_{m_2} where $||c_{m_1}, c_{m_2}||_2 < d_{\text{th}}$
    if neighbors found
      add both neighbors and the query centroid to a *stairs* set

(a) *stairs* sequence (b) *dumpster* sequence

Fig. 12.14 Semantically annotated polygonal models for two datasets (top image is birds-eye view and the bottom image is the close-up view). (a) *stairs* ($\approx$ 2000 frames) (b) *dumpster* ($\approx$ 3000 frames).

Most of the above surface class labels are assigned directly while performing the region growing operations through dot products and distance computations. The only exception to this rule is the *stairs* class, which is treated separately. Algorithm 12.2 presents the individual steps that lead to the labeling of a polygon or group of polygons as being part of the *stairs* class.

The d_{th} threshold dictates the desired search space for each candidate stair step. In Algorithm 12.2, it describes the maximum radius of a sphere containing the nearest neighbors (from c_{m_2}) of each horizontal centroid in c_{m_1}. Setting the parameter to a high value will render the algorithm to search for more neighbors, which might bring an additional computation burden. An implementation example is to set the parameter to 0.2m, as this provides an intuitive way on restricting a stair step size, while at the same time giving good results for almost all datasets tested against.

12.1.6 3D Mapping Performance

To evaluate the system's performance, all its components have been carefully monitored both under stressful conditions as well as during idle operations. The computation time and memory usage of each subsystem of the 3D mapping core are particularly interesting with respect to the real-time performance capabilities of the mapping architecture. The evaluation of the mapping pipeline includes individual results for the following components:

- *Stereo + VO* - produces the disparity image from the stereo system and computes the global pose estimate;
- *PCD→LocalOctree* - produces the local point cloud dataset (PCD) from the disparity image and creates/updates the local octree;
- *GlobalOctree growing* - expands the local octree's boundaries when needed;
- *LocalOctree→Polygon→GlobalOctree* - generates the polygonal model and merges it with the existing model;
- *GlobalOctree→GlobalModel* - refines the model (polygonal merging, polygonal growing, outlier removal);
- *Semantic (Functional) Labeling* - annotates the model.

Figure 12.15 presents results for the datasets that correspond to some of the models presented in Figure 12.6. The majority of the model-building components have low computational requirements, and almost consistently, *Stereo + VO* requires the majority of resources. To address this, the *Stereo + VO* component has been implemented in a separate process from the other (model-building) components. By concurrently running the two processes, the global map can be updated every 250ms on average. An additional performance gain could be achieved by executing the stereo computations offline, in-camera [Kon97].

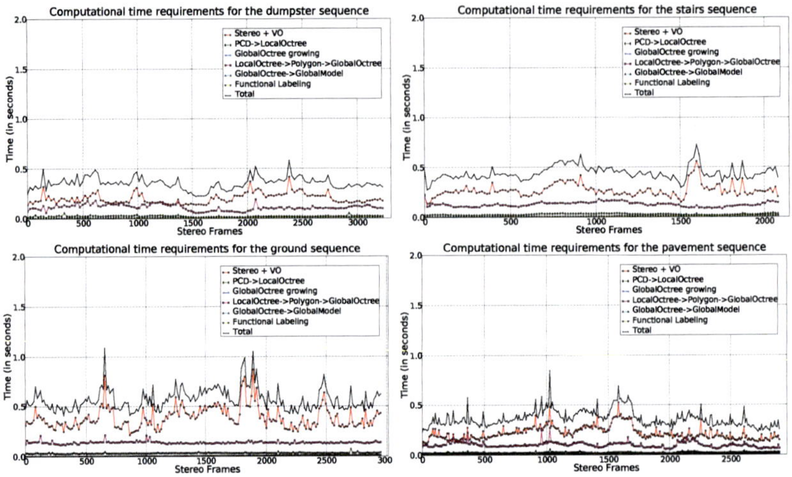

Fig. 12.15 Computational time of each 3D mapping component for several tested datasets (see Figure 12.6)

Figure 12.16 shows the memory usage plot for the same datasets. To illustrate the system's operation better, the amount of memory allocated at each frame has been analyzed, both when new data is received and has to be integrated into the model, as well as after the model has been estimated and the initially allocated memory has been freed. Therefore, the two curves, blue and red, show the quantity of memory occupied before $\mathcal{M}_L^{(t)}$ is created, and after.

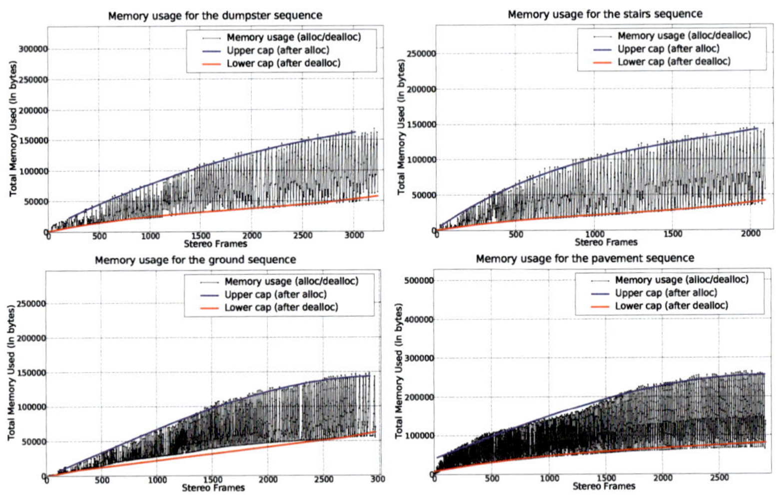

Fig. 12.16 Memory usage for the creation of the final polygonal models $\mathcal{M}_F^{(t)}$ for several tested datasets (see Figure 12.6)

The presented results indicate the favorable computational performance of the mapping approach proposed by Leaving Flatland. The implementation of the system was tested on a regular dual-core Centrino notebook attached to the RHex robot as seen in Figure 12.4. Due to the parallelization capabilities of the octree-enabled mapping architecture, preliminary results on quad-core systems have shown even more performance gains. Moving stereo computations in the camera would potentially enable the map building process to run in less than 50ms on modern systems, thus freeing more resources for other tasks such as the motion planner, or higher level executives.

12.2 Semantic Map Usage and Applications

The semantic maps presented in the previous sections exhibit some of the capabilities of the Leaving Flatland mapping architecture. The following sections present usage examples.

12.2.1 Hybrid Model Visualizations

Though extremely efficient from an architectural point of view, a negative aspect of convex planar patches based maps is that they are not the most suitable surface representations for visualization. Because one of the possible applications of Leaving Flatland is to enable remote visualization of a scene, like for instance, by a teleoperator, a more suitable representation for visualization has to be implemented. A viable solution is to represent the estimated world model as a *hybrid* concept, with polygonal patches for certain parts of the environment, and sampled point cloud data for the rest. This would enable the transmission of the model over a bandwidth-limited network, and also provide a visually rich as well as schematically coherent views for the remote operator. The reasons for selecting the above hybrid representation are twofold:

- While the full point cloud data is rich in terms of visualization, it requires a large bandwidth for transmission, whereas the polygonal model alone is a compact representation but is not suitable for parts of the scene that are more complex, like vegetation or surfaces exhibiting a lot of variations in curvature;
- Since some of the proposed class labels, such as *vertical*, provide a binary decision with respect to the property of traversability (*i.e.*, either allowed to traverse, but needs a special controller/trajectory, or cannot traverse at all), the respective polygons could easily be presented using the original point cloud data to the remote operator, while still retaining the respective class label property; this has the added advantage that the robot will not be able to navigate over these parts of the world, but at the same time the operator would visualize their structure and infer additional information regarding the robot's position, next possible task, etc.

Leaving Flatland thus includes a scheme in which parts of the world are represented using either the polygonal model or the sampled point cloud data in order to better visualize the scene in an efficient manner. An example of such view of the hybrid model is presented in Figure 12.17. The ground and level cells are not extremely interesting for visualization and are therefore represented using simplified polygonal patches (colored in green). The rest of the world, including vertical structures and stairs, is represented with points for better visualization.

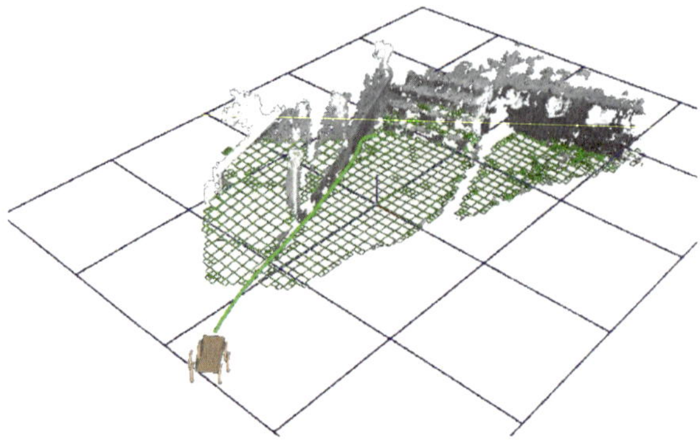

Fig. 12.17 Hybrid model visualized remotely. Ground and level cells are plotted as polygons and the rest as points.

12.2.2 *Motion Planning for Navigation*

The motion-planning subsystem is called periodically to generate a trajectory that connects the current position of the robot to a user-designated target. The planner is given a semantic map of the environment built by the 3D perception system. Its output is a trajectory that consists of a sequence of gait primitives to be executed by the robot, and their intermediate milestones. A trajectory with a low (but not necessarily optimal) execution cost is seeked. The execution cost is defined as a function of local terrain characteristics such as slopes and unevenness, as in [SDW93].

To achieve real-time motion planning, the 3D semantic map created by the perception system is represented as a topological graph (see Figure 12.18). Each contiguous collection of polygonal patches with the same semantic label is represented

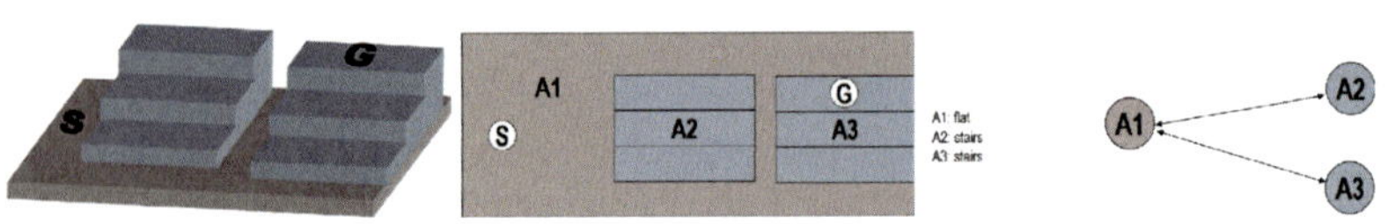

Fig. 12.18 The 3D semantic map model represented as a topological graph with nodes representing contiguous surface regions with the same semantic label and transitions

as a node in the graph, while the graph edges represent transitions between different world regions. This decomposition of the 3D model into regions that are locally 2D allows the motion planning system to use specific planners for each region, and greatly simplifies terrain traversability and collision checking.

The motion planning system implemented in Leaving Flatland is thus comprised of a multi-region planner and a repertoire of basic primitive planners that work only on a certain semantic type of terrain. The following types of primitive planners are used, classified by the type of regions on which they are applicable:

- *unlabeled terrain*: A* search on a discretized (x, y, θ) state space [SDW93];
- *flat ground*: sample-based SBL motion planner [SL02], which builds trees bidirectionally in configuration space to connect the start and goal;
- *stairs* or *steep slopes*: the robot's heading is restricted to either face up or down the slope, and the stair-climbing gait primitive is used; the planner examines a 1D line in configuration space for collisions;

The unlabeled terrain planner is also used as a backup when no other primitive planner is applicable. Note that the unlabeled terrain primitive planner can be applied to any mode for which any other primitive planners can be applied.

Each primitive planner can only be applied to modes whose regions are connected and do not overlap in the x-y plane (otherwise, this is considered an invalid mode). The multi-region planner selects (valid) modes, initial and final configurations for primitive planner queries, and chooses the fastest planner for the given query. The implementation details together with a comprehensive discussion regarding its scope is given in [RSM$^+$09].

The motion planning system was tested with and without semantic annotations in the 3D map. For a flat terrain composed of over 20,000 triangles, with a point target 5m away from the robot's initial position, the decomposition step took approximately 220ms. If the terrain is unlabeled, the multi-region planning stage took 184ms to generate a path (averaged over 10 runs). Labeled as flat ground, it took 53ms. A two-level problem was tested consisting of flat ground and stairs, with a point target on a second level, approximately 2m above the robot. The decomposition step took approximately 200ms. Multi-region planning took 886ms on unlabeled terrain, and 134ms on labeled terrain. Similar increases in computational speed were observed in several other experiments.

Experiments on terrain generated from the 3D mapping module (*e.g.* Fig. 12.19) show that the planner can generally produce a path to targets up to approximately 5m away from the robot's initial position in less than a second. The running time can vary significantly depending on the size of the model, the shape of obstacles, and the performance of heuristics. However, due to the highly efficient decomposition scheme, the motion planner usually achieves fast planning times ranging from 0.05s-0.1s for SBL, to 0.5s-2s for A*, and almost instant computations ($\approx$0s) for the 1D straight line stairs/slopes planner.

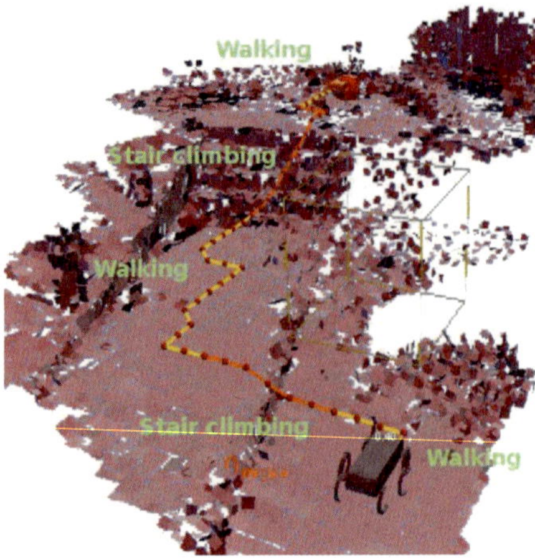

Fig. 12.19 Motion planning example based on semantic polygonal labels for the *dumpster* dataset

12.3 Summary

This chapter presented the theoretical and implementation details of Leaving Flatland, a comprehensive system for real-time 3D mapping and planning for general purpose mobile robots, with an emphasis on the RHex robot.

The proposed system includes mapping and localization techniques, coupled with higher level polygonal modeling annotations for stereo data. These annotations result in heuristic semantic labels for different terrain types, which are extremely helpful for motion planners to choose the most appropriate gaits for the robot. The motion planner used for the experiments presented herein, utilizes a new decomposition technique to convert a 3D model locally into 2D regions, and builds upon a repertoire of planning methods specialized for specific terrain types, such as flat ground and stairs.

The performance of the mapping system both in terms of the quality of the modeling and the processing speed, enables the deployment of the proposed architecture on real robots, and is suitable for fast, real-time 3D navigation.

Chapter 13
Conclusion

"A conclusion is the place where you get tired of thinking."

ARTHUR BLOCH

As autonomous personal robotics systems come to age, they need to find ways of functioning in human living environments, that is, navigating and interacting with the surrounding world in an effective manner. This means a robot should be capable of going anywhere where it can physically fit, find energy sources and recharge its batteries when their capacity is running low, and perform useful chores such as cleaning tables. These behaviors require support from complex perception routines which can recognize power plugs, door handles, and certain structures and objects in the world. Therefore one of the most important aspects deciding on the applicability of personal robotics in real world environments is represented by the robot's ability to perceive and understand the semantic meanings of the objects present in the world. The key challenge in achieving that is being able to build semantic 3D perception architectures that can enable complex mobile manipulation scenarios.

This book proposed Semantic 3D Object Maps as a novel representation that satisfies these requirements, and showed how these maps can be automatically acquired from dense 3D range data. The main results of this investigation are a set of theoretical and methodological contributions, fully implemented and tested on several mobile manipulation platforms in realistic scenarios.

The first part of the book covered the Semantic 3D Object Mapping kernel. In particular, the problem of data acquisition and map representation has been tackled in Chapter 2, while in Chapter 3 we have presented and discussed the major requirements of a system architecture for Semantic 3D Object Maps. Each of the individual processing steps presented in the architecture have been fully implemented and described in the subsequent Chapters 4-9.

In Chapter 4 we have presented important contributions to the problem of informative 3D point feature representations from noisy point cloud data. Both our Point Feature Histograms (PFH) and Fast Point Feature Histograms (FPFH) representations have shown to be discriminative features and gave very good results when used in applications such as point correspondence searches for registration (Chapter 5) and surface classification with respect to the underlying geometric surface primitive that the point is lying on (Chapter 8). To reduce the complexity for point

R.B. Rusu: *Semantic 3D Object Maps for Everyday Robot Manipulation*, STAR 85, pp. 197–200.
DOI: 10.1007/978-3-642-35479-3_13

cloud registration, we have shown methods to analyze the uniqueness and persistence of the proposed feature representations using different distance metrics over multiple scales, resulting in more compact subsets of points which characterize and represent the larger input datasets.

Chapter 5 presented two coarse registration methods using our feature representations that bring datasets acquired from different arbitrary poses into the same coordinate system, thus creating a consistent global point cloud model. We have applied the proposed registration methods to a variety of datasets and have shown them to be robust in the presence of different initial starting solutions. To smooth out overlapping areas in the resultant compact point clouds, but also to remove noise and fill holes in the data, we have presented a Robust Moving Least Squares algorithm that can resample data and preserve fine details useful for geometric processing steps such as point cloud segmentation.

To split large quantities of data into smaller chunks for faster processing, Chapter 6 has tackled different clustering and segmentation techniques which can be used to group points with the same geometric information together, or to segment and simplify resultant maps using primitive 3D geometries. These concepts have been applied thoroughly in Chapter 7 for the problem of creating complete 360° semantic models of the world, with an emphasis for kitchen environments which exhibit structural elements such as cupboards, tables, and kitchen appliances. To support 3D collision detection and path planning routines for safe navigation and manipulation of objects, we have shown ways of creating decoupled Triangulated Surface Maps, that represent the underlying sampled surfaces with a greater accuracy than point cloud models.

In Chapter 8 we presented novel scene interpretation mechanisms for point cloud data acquired using noisy sensors such as active stereo. The surface geometry around a point was characterized using our discriminative point feature representations (PFH and FPFH), which have been shown to give very good results for the problem of point classification with respect to the underlying geometric surface primitive the point is lying on. By carefully defining 3D shape primitives, and making use of powerful machine learning classifiers, our framework was able to robustly segment point cloud scenes into geometric surface classes.

Additionally, we have also tackled the problem of object categorization, by proposing a novel global descriptor (Global Fast Point Feature Histograms) that makes use of the previously acquired local annotations from a FPFH based classification system. In Chapter 9 we proposed object reconstruction mechanisms that represent the objects to be manipulated using hybrid shape-surface geometric models. These models were acquired out of sensed point cloud data using robust shape segmentation and surface triangulation methods, and we showed methods that could fit primitive geometric shapes to the data in order to reconstruct and infer missing information.

The contributions presented in this book have been evaluated as parts of complete mobile manipulation systems in the context of three distinct application scenarios. The first application was presented in Chapter 10 and relates to the problem of cleaning tables by moving the objects present on them with personal robotic assistants operating in dynamic environments. The proposed architecture constructed 3D dynamic collision maps, and annotated the surrounding world with semantic labels. To support a simple grasping module, object clusters supported by tables have been extracted with real-time performance. The second application presented in Chapter 11 showed a novel architecture for door and handle identification from noisy 3D point cloud maps acquired using tilting laser sensors, with excellent applicability for mobile manipulation platforms operating in indoor environments. A third complete application was shown in Chapter 12 for the problem of real-time semantic mapping from stereo data for efficient navigation.

All the algorithms and concepts presented in this book have been implemented and tested on data acquired in realistic everyday human living environments. The experimental results were carried out on numerous mobile manipulation platforms, including the TumBot (see Figure 2.2 left) mobile manipulation platform at the Technische Universität München, the RHex (see Figure 12.4) mobile robot at SRI, and the PR2 (see Figure 10.1) personal robotic assistant at Willow Garage.

To stimulate and foster the development of novel research ideas, we are making the entire software infrastructure used in in the experimental part of the book available as part of the ROS (Robot Operating System) open source software initiative. We encourage researchers working in different fields to do the same.

Despite the encouraging results presented in this book, there are a list of open issues that still remain to be tackled for future research:

- Though tested on a series of indoor scenes, the methods and algorithms presented in this work need to be tested on an even larger database of datasets representing human living environments. To achieve this, we acknowledge the need of a repository of processing algorithms, each working extremely well under particularly constrained conditions. The goal would be then to build higher level task executives that can automatically select the best algorithm for a certain problem given some input conditions. Additionally, multiple processing methods could be started in parallel (or sequentially if the application time constraints are relaxed) and the executive could select the overall results based on the individual results of each method.

  We are currently already taking steps towards solving this problem, by incorporating a large variety of different algorithms in the ROS initiative. However, depending on the complexity of the problem to be solved, we may need to rethink this approach in the future.
- In this book we already tackled the problem of environment dynamics, but mostly in terms of building fast dynamic obstacle maps for the problem of collision avoidance with a mobile manipulation platform. An important aspect that we have not modelled at all is the tracking of objects or surfaces while they are being moved or manipulated. To solve this however, we envision two different approaches:

  - the use of faster 3D sensing devices that could record maps at a higher frame rate, together with the optimization of our processing routines even more to be able to extract meaningful information in application with tight computational constraints;

- the fusion of 2D camera images within the 3D mapping system, and the use of dedicated image processing algorithms which already present very good results for tracking movable targets.

- Though we hopefully proved the need for 3D geometry in the context of acquiring semantic maps of indoor environments, this book has not addressed the fusion of texture information with geometry at all. Texture becomes extremely important for particular applications which geometry alone cannot solve. For example, the shape of a bottle of mustard and a bottle of ketchup could be identical, but their uses are different, so an intelligent robot assistant should use color (yellow vs. red) to differentiate between them.

Though these are important issues that need to be solved in future research initiatives, we are confident that in this book we managed to take small steps towards the realization of semantic mapping systems, which we believe will be the cornerstone of all mobile manipulation platforms in the future.

Appendix A
3D Geometry Primer

Following the notations presented in the list of symbols (page XXIII), the most primitive data representation unit in a general n-dimensional space $\mathbb{S}^n$ is represented by a point $\boldsymbol{p}^n = (x_1, x_2, \cdots x_n)$, with n being the number of tuples or dimensions. This book most often makes use of the two-dimensional and three-dimensional Euclidean spaces $\mathbb{E}^2$ and $\mathbb{E}^3$, though higher order dimensional spaces are also used. Therefore, an unanimous notation for a 2D point will be set from now on to $\boldsymbol{p}^2$, while a 3D point will be expressed as $\boldsymbol{p}^3$ or $\boldsymbol{p}$ for simplicity.

A.1 Euclidean Geometry and Coordinate Systems

Without any exaggeration, Euclidean geometry and the associated Euclidean space are two of the most used concepts in classical mathematics, as until the 19th century *Euclidean* was a synonym of the word *geometry*. Ever since though, a number of non-Euclidean geometries have been proposed, which in essence contradict the earlier postulates of Euclidean geometry. From these, the most important models are represented by *hyperbolic geometry* and *elliptic geometry*. The latter includes a simplified formulation, the *spherical geometry*, which is interesting for some of the concepts presented in this book.

In the Euclidean $\mathbb{E}^3$ space, which is considered infinite [New56], the given coordinates of a point $\boldsymbol{p}$ are specified with respect to a fixed origin, in a given coordinate system frame. Two of the most commonly used coordinate systems in $\mathbb{E}^3$, are the Cartesian coordinate system and the spherical coordinate system, both of equally significant importance for 3D perception systems. Each coordinate system in E^3 has three axes perpendicular to each other and selected as $\vec{x}$, $\vec{y}$, and $\vec{z}$ as a naming convention, with a length of one unit. For a point with real coordinates $\boldsymbol{p} \in \mathbb{R}^3$, its coordinates in the Cartesian coordinate system can be written as follows:

$$\boldsymbol{p} = [x, y, z]^T = \begin{bmatrix} x \\ y \\ z \end{bmatrix} \tag{A.1}$$

$$x, y, z \in \mathbb{R}$$

The three coordinates x, y, z represent the number of units (distance) on each of the three axes from the origin of the coordinate system.

Besides points, Euclidean geometry defines the notion of a *direction vector*. The direction of movement from a point $\boldsymbol{p}_i$ to another point $\boldsymbol{p}_j$ is defined as follows:

$$\mathsf{v} = \boldsymbol{p}_j - \boldsymbol{p}_i = [x_j, y_j, z_j]^T - [x_i, y_i, z_i]^T = [x_j - x_i, y_j - y_i, z_j - z_i]^T \quad \text{(A.2)}$$

To formulate concepts such as the distance between two points $\boldsymbol{p}_i$ and $\boldsymbol{p}_j$, Euclidean geometry makes use of an operation called *dot product* (or *inner product*), defined for $\mathbb{E}^3$ as follows:

$$\langle \boldsymbol{p}_i, \boldsymbol{p}_j \rangle = \boldsymbol{p}_i \cdot \boldsymbol{p}_j = \sum_{n=1}^{3} \boldsymbol{p}_i \boldsymbol{p}_j = x_i x_j + y_i y_j + z_i z_j \in \mathbb{R} \quad \text{(A.3)}$$

In a spherical coordinate system, a vector can be written as:

$$\boldsymbol{p} = [r, \theta, \psi]^T = \begin{bmatrix} r \\ \theta \\ \psi \end{bmatrix} \quad \text{(A.4)}$$
$$r \in \mathbb{R}^+, \quad 0^\circ \leq \theta \leq 180^\circ, \quad 0^\circ \leq \psi \leq 360^\circ$$

The first coordinate r represents the radial distance of the point, and can be expressed as the Euclidean distance (see Section A.2) from the origin of the coordinate system to the point $\boldsymbol{p}$. The second coordinate θ represents the angle between the line from the origin of the coordinate system to $\boldsymbol{p}$ and the zenith direction. Finally, the third coordinate ψ represents the angle between the reference direction on the chosen plane and the line from the origin of the coordinate system to the projection of $\boldsymbol{p}$ on the plane.

A.2 Distance Metrics

The following describes some of the distance metrics most used throughout this book.

The Manhattan (L1) distance:

$$d_{L1} = \sum_{i=1}^{n} |p_i - q_i| \quad \text{(A.5)}$$

The Euclidean (L2) distance:

$$d_{L2} = \sqrt{\sum_{i=1}^{n} (p_i - q_i)^2} \quad \text{(A.6)}$$

The Jeffries-Matusita (JM) metric (also known as Hellinger distance) is similar to the L2 (Euclidean) norm, but more sensitive to differences:

$$d_{JM} = \sqrt{\sum_{i=1}^{n} (\sqrt{p_i} - \sqrt{q_i})^2} \quad \text{(A.7)}$$

The Bhattacharyya distance is widely used in statistics to measure the statistical separability:

$$d_B = -ln\sum_{i=1}^{n} \sqrt{p_i - q_i} \tag{A.8}$$

Two popular measures for histogram matching in literature, the Chi-Square (χ^2) divergence and the Kullback-Leibler (KL) divergence, are defined as:

$$d_{\chi^2} = \sum_{i=1}^{n} \frac{(p_i - q_i)^2}{p_i + q_i} \tag{A.9}$$

$$d_{KL} = \sum_{i=1}^{n} (p_i - q_i) \cdot ln\frac{p_i}{q_i} \tag{A.10}$$

The Histogram Intersection Kernel (HIK):

$$d_{HIK} = \sum_{i=1}^{n} \min(p_i, q_i) \tag{A.11}$$

A.3 Geometric Shapes

The equation of a line is:

$$y = p + vt \tag{A.12}$$

where $\boldsymbol{p} = [x_0, y_0, z_0]^T$ is a point on the line, and $\boldsymbol{v} = [a, b, c]^T$ is a direction vector.

The equation of a plane is:

$$ax + by + cz + d = 0 \tag{A.13}$$

where $\vec{\boldsymbol{n}} = [a, b, c]^T$ is the normal of the plane.

The equation of a sphere is:

$$(x - x_c)^2 + (y - y_c)^2 + (z - z_c)^2 = r^2 \tag{A.14}$$

where $\boldsymbol{p}_c = [x_c, y_c, z_c]^T$ represents the center of the sphere, and r is the radius of the sphere.

The equation of a right cone with the vertex pointing up and the base located at $\vec{z} = 0$ is:

$$\frac{x^2 + y^2}{c^2} = (z - z_0)^2, \; c = \frac{r}{h}, \; z_0 = h \tag{A.15}$$

where h is the height of the cone, and r is the radius of the base oriented along the $\vec{z}$ axis.

The equation of torus radially symmetric about the $\vec{z}$ axis is:

$$(R - \sqrt{x^2 + y^2})^2 + z^2 = r^2 \tag{A.16}$$

where R is the distance from the center of the tube to the center of the torus, and r is the radius of the tube.

The equation of an elliptic cylinder is:

$$\left(\frac{x}{a}\right)^2 + \left(\frac{y}{b}\right)^2 = 1 \tag{A.17}$$

where a and b are the semimajor and semiminor axes.

Appendix B
Sample Consensus

The RANdom SAmple Consensus (RANSAC) [FB81] principle states the following. If ϵ is the probability of picking a sample that produces a bad estimate (*i.e.* outlier), then $1-\epsilon$ is the probability of picking at least one good sample (*i.e.* inlier). This means that the probability of picking s good samples becomes $(1-\epsilon)^s$. For k trials, the probability of failure becomes $(1-(1-\epsilon)^s)^k$. If p is the desired probability of success (*e.g.* $p = 0.99$), then:

$$1 - p = (1-(1-\epsilon)^s)^k \quad \Rightarrow \quad k = \frac{\log(1-p)}{\log(1-(1-\epsilon)^s)} \tag{B.1}$$

The RANSAC cost function is defined as:

$$C_{\mathrm{r}} = \sum_i \rho_{\mathrm{r}}(e_i^2), \qquad \rho_{\mathrm{r}}(e^2) = \begin{cases} 0, & e^2 < T^2 \\ \text{constant}, & e^2 \geq T^2 \end{cases} \tag{B.2}$$

while the MSAC (M-estimator SAmple Consensus) [TZ00] cost function is defined as:

$$C_{\mathrm{m}} = \sum_i \rho_{\mathrm{m}}(e_i^2), \qquad \rho_{\mathrm{m}}(e^2) = \begin{cases} e^2, & e^2 < T^2 \\ T^2, & e^2 \geq T^2 \end{cases} \tag{B.3}$$

where T is a user given threshold for considering inliers.

Appendix C
Machine Learning

Without a doubt, the last decade has seen an incredible progress in the research fields of machine learning and pattern recognition in data. This progress is correlated with the advancements in computational technology, which enable algorithms to run faster than they did years ago, but also with better problem formulations and classification algorithms which slowly made their way and established themselves as practical solutions for a variety of problems in robotics.

Due to space constraints, it would be impossible to cover all the popular machine learning methods used today, in this book. The reader is encouraged to search for general textbooks on the subject, such as [Bis06, DHS00, Mit97] to name a few. However, this chapter will cover some of the basics of two of the most used learning methods in the context of this book, namely Support Vector Machines (SVM) and Conditional Random Fields (CRF).

Before diving into the specifics of each method, it is important to mention that the machine learning community has divided the learning methods into three distinct categories, based on the problems they are used for [Bis06]: a) supervised learning; b) unsupervised learning; and c) reinforcement learning. The first category of methods attempts to learn a mapping between sets of data comprising both the input and the desired output of a particular problem. Once the mapping is learned and a model is obtained, all the subsequent test data will be validated against this model, in a classification or a regression framework. The second category makes use of clustering or density estimation techniques for example, to learn an internal model from a set of training data without having an explicit output target. Finally the third method creates models that adapt themselves over time based on the actions taken to maximize a certain reward function with respect to a given problem. The two methods mentioned above (SVM and CRF) fall into the first category of supervised learning methods.

C.1 Support Vector Machines

Support Vector Machines (SVM) are part of a distinct family of supervised learning algorithms called kernel-based methods, where data predictions are based on linear combinations of a kernel function evaluated at the training data points. Originally developed in the context of optical character recognition applications

[BGV92, CV95], SVM methods have become popular for solving many different other problems in classification and regression frameworks. Though at their core they are binary classifiers, meaning that they are only capable of separating between two different classes of data, many extensions have been developed in the context of multi-class classification.

The determination of the SVM model parameters corresponds to a convex optimization problem, which leads to global optimum solutions. In its simplest form, the SVM formulation for a set of training vectors $\boldsymbol{x}_i \in \mathbb{R}^n$ and a binary set of classes $c_{1,2}$ can be given as follows. Let the desired output $\boldsymbol{y}_i$ be:

$$y_i = \begin{cases} 1, \text{ for } \boldsymbol{x}_i \in c_1 \\ -1, \text{ for } \boldsymbol{x}_i \in c_2 \end{cases} \tag{C.1}$$

Then, let the two classes c_1 and c_2 be linearly separable by a hyperplane (see Figure C.1) described as:

$$\mathbf{w}^T \boldsymbol{x} + b = 0 \tag{C.2}$$

where $\mathbf{w}$ is the hyperplane's normal, b is a bias parameter, and $\frac{-b}{|\mathbf{w}|}$ the perpendicular distance from the hyperplane to the origin. This means that there exists at least one choice of $\mathbf{w}$ and b such that:

$$\begin{cases} \mathbf{w}^T \boldsymbol{x}_i + b > 0, \text{ for } \boldsymbol{y}_i = 1 \\ \mathbf{w}^T \boldsymbol{x}_i + b < 0, \text{ for } \boldsymbol{y}_i = -1 \end{cases} \tag{C.3}$$

Therefore the decision function becomes $\mathcal{F} = sign(\mathbf{w}^T \boldsymbol{x} + b)$ with $\boldsymbol{x}$ as test data. A Support Vector Machine classifier attempts to find the variables $\mathbf{w}$ and b such that the smallest distance from the hyperplane to any of the samples – called the *margin*, is maximized. By further describing the constraints as:

$$\begin{cases} \mathbf{w}^T \boldsymbol{x}_i + b \geq d_+, \text{ for } \boldsymbol{y}_i = 1 \\ \mathbf{w}^T \boldsymbol{x}_i + b \leq d_-, \text{ for } \boldsymbol{y}_i = -1 \end{cases} \tag{C.4}$$

where d_+ and d_- represent the outer boundary. The maximal distance between $\mathbf{w}^T \boldsymbol{x}_i + b$ is ± 1. By selecting $\mathbf{w}$ and b with the maximal margin, $d_+ = 1$ and $d_- = 1$, the margin becomes:

$$d_+ + d_- = 2 \tag{C.5}$$

By combining Equation C.4 and C.5, we obtain:

$$\boldsymbol{y}_i(\mathbf{w}^T \boldsymbol{x}_i + b) \geq 1 \tag{C.6}$$

The margin is determined by the points that lie closest to the hyperplane, and therefore maximizing the margin means building a better separation boundary so that the classification error is minimal. The subset of data points $\boldsymbol{x}_i$ lying on this boundary are also known as the support vectors of the classification. In the example in Figure C.1 the margin is $\frac{2}{\|\mathbf{w}\|}$. Therefore, to find the maximum margin, $\|\mathbf{w}\|$ has to be minimized, which in turn is equivalent to minimizing $\|\mathbf{w}\|^2$.

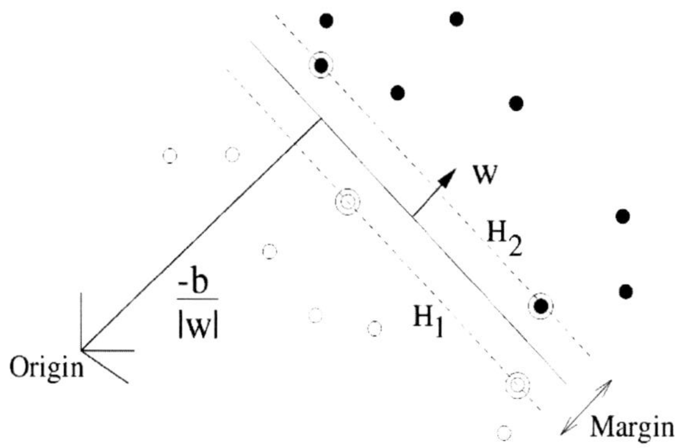

Fig. C.1 A theoretical hyperplane dividing two linearly separable classes (white points - c_1, black points - c_2). The data points colored in red represent the support vectors of the hyperplane.

This problem can formally be written as a convex optimization problem:

$$\begin{aligned} &\min \frac{1}{2}\|\mathbf{w}\|^2 \\ \text{for:}\quad & y_i(\mathbf{w}^T x_i + b) \geq 1 \end{aligned} \tag{C.7}$$

This constrained optimization problem can be solved better in a Lagrangian formulation [Bur98], as the constraints are easier to handle and the training data will only appear in the form of dot products between vectors. By introducing the Lagrangian multipliers $\alpha_i \geq 0$, the Lagrangian function is obtained as:

$$L_P = \frac{1}{2}\|\mathbf{w}\|^2 - \sum_i \alpha_i y_i(\mathbf{w}^T x_i + b) + \sum_i \alpha_i \tag{C.8}$$

The minimization with respect to $\mathbf{w}$ and b means setting the derivatives of L_P for $\mathbf{w}$ and b to 0. Substituting back results in:

$$\begin{cases} \mathbf{w} = \sum_i \alpha_i y_i x_i \\ 0 = \sum_i \alpha_i y_i \end{cases} \Rightarrow L_D = \sum_i \alpha_i - \frac{1}{2}\sum_{i,j} \alpha_i \alpha_j y_i y_j k(x_i, x_j) \tag{C.9}$$

L_D is referred to as the dual form of the primary L_P, and its constraints differ from the ones of L_P [Bis06]. The term $k(x_i, x_j)$ is a kernel function which maps the linear non-separable data to a higher dimension where a linear separation by a hyperplane is possible. Any function that satisfies the following conditions can be a kernel:

1. The function k has to be symmetric: $k(x_i, x_j) = k(x_j, x_i)$;
2. The kernel matrix $K_{i,j} = k(x_i, x_j)$ needs to be positive semi-definite for all points x.

A good reference on engineering kernel functions is given in [STC04]. Some of the most commonly applied kernels are given below:

- linear kernel: $k(\boldsymbol{x}_i, \boldsymbol{x}_j) = \boldsymbol{x}_i^T \boldsymbol{x}_j$;
- polynomial kernel: $k(\boldsymbol{x}_i, \boldsymbol{x}_j) = (\boldsymbol{x}_i^T \boldsymbol{x}_j + \delta)^d$, with $\delta > 0$;
- sigmoidal kernel: $k(\boldsymbol{x}_i, \boldsymbol{x}_j) = tanh(\kappa \boldsymbol{x}_i^T \boldsymbol{x}_j + \delta)$, with $\kappa > 0, \delta < 0$;
- Radial Basis Function (RBF) kernel: $k(\boldsymbol{x}_i, \boldsymbol{x}_j) = \exp^{-\gamma \cdot d(\boldsymbol{x}_i, \boldsymbol{x}_j)}$ with $\gamma > 0$, where d can take the form of certain distance metrics, including:
  - $d(\boldsymbol{x}_i, \boldsymbol{x}_j) = \sum_k |\boldsymbol{x}_{i,k} - \boldsymbol{x}_{j,k}|^{\frac{1}{2}}$ for the Sublinear RBF kernel;
  - $d(\boldsymbol{x}_i, \boldsymbol{x}_j) = \sum_k |\boldsymbol{x}_{i,k} - \boldsymbol{x}_{j,k}|^1$ for the Laplacian RBF kernel;
  - $d(\boldsymbol{x}_i, \boldsymbol{x}_j) = \sum_k |\boldsymbol{x}_{i,k} - \boldsymbol{x}_{j,k}|^2$ for the Gaussian RBF kernel.

As previously mentioned, SVMs are binary classifiers. Though native multi-class SVM solutions exist, they can be computationally expensive. In order to keep the classification process fast and still retain the ability to separate multiple classes, a simple method called *one-against-all* is used, where one binary SVM is used for each class to separate members of that class from members of other classes.

C.2 Conditional Random Fields

Conditional Random Fields belong to a more general family of probabilistic graphical models, along other formulations such as Bayesian Networks or Hidden Markov Models, which use solutions from both the probability theory and graph theory to solve learning problems. A probabilistic graphical system borrows the concepts of nodes and edges (links) from graph theory, and provides an intuitive interface for modeling variables and their inter-dependencies. Each node represents a random variable and each link is used to interconnect two nodes to symbolize their probabilistic dependency to each other. There are two basic types of probabilistic graphical models:

- *directed* - used to model causal relationships between random variables;
- *undirected* - better suited for expressing soft constraints between random variables.

Directed graphs can be converted into undirected graphs using a technique called moralization where the parents of each node, if they exist, get connected by an undirected edge and the remaining directed edge gets replaced by undirected edges [Bis06]. In general the opposite conversion is less common and more problematic. Additionally, there are undirected graphs whose conditional independence properties cannot be expressed in a directed graph with the same variables and vice versa.

To solve inference problems in a probabilistic manner, it is often useful to convert directed and undirected graphs into so-called factor graphs [FKaLW98]. A factor graph is a bipartite graph that expresses the structure of the factorization as:

$$p(\boldsymbol{x}) = \prod_j f_j(\boldsymbol{x}_j). \tag{C.10}$$

where x_j is a subset of the variables. A factor graph has a variable node for each variable $\boldsymbol{x}_j$, a factor node for each local function f_j, and an edge-connecting variable node $\boldsymbol{x}_j$ to factor node f_j if and only if $\boldsymbol{x}_j$ is an argument of f_j [Bis06].

In general there are two distinct approaches to solve decision problems, using:

- *generative models*, which model the distributions of input variables $\boldsymbol{x}$ and output variables $\boldsymbol{y}$. The models are called generative, because they have the ability to both infer from the input variables to the output class and create values for the input space by sampling over the output. Generative models need to model the joint distribution $p(\boldsymbol{x},\boldsymbol{y})$;
- *discriminative models*, which only model the conditional probability distribution $p(\boldsymbol{y}|\boldsymbol{x})$ by inferring over the input variables $\boldsymbol{x}$ (called observations), according to the classifier characteristics of this approach. Each new observation $\boldsymbol{x}$ gets assigned to a class $\boldsymbol{y}$.

Generative models are obviously more complex to handle than discriminative ones, because the joint probability for both variables $\boldsymbol{x}$ and $\boldsymbol{y}$ has to be found. Furthermore the number of observations $\boldsymbol{x}$ in many applications can be high, so the modeling and the inference become intractably complex.

Conditional Random Fields are undirected discriminative graphical models, where the distribution of each discrete random variable $\boldsymbol{y}$ in the graph is conditioned on an input sequence $\boldsymbol{x}$, thus representing a discriminative probability $p(\boldsymbol{y}|\boldsymbol{x})$ [LMP01].

A convenient representation when modeling with undirected graphical models is given by the concept of *cliques*. A clique C is a fully connected subset of the nodes in a graph such that each node in the clique is connected by an edge to all the other nodes in the clique. This can be expressed as:

$$\forall n_i \in V_C : \exists n_j \in V_C : (n_i, n_j) \in E_C,\ V_C \subseteq V, E_C \subseteq E \tag{C.11}$$

where E is the set of edges, V is the set of nodes in the graph, E_C is the set of edges and V_C the set of nodes in the clique C. A *maximal clique* is a clique that cannot include any other nodes from the graph without it ceasing to be a clique [Bis06].

A definition of Conditional Random Fields can be given as follows [LMP01]. Let $G = (V,E)$ be an undirected graph over the set of random variables Y and X, with V representing the set of nodes, and E the set of edges in G. Let $Y = (Y_v)_{v \in V}$, so that Y is indexed by the vertices of G. Then (Y,X) is a X conditional random field, if the random variables Y_v obey the Markov property with respect to G:

$$p(Y_v|X, Y_w, w \neq v) = p(Y_v|X, Y_w, w \sim v) \tag{C.12}$$

where $w \sim v$ means that w and v are neighbors in G.

The conditional probability $p(\boldsymbol{y}|\boldsymbol{x})$ can be written as:

$$p(\boldsymbol{y}|\boldsymbol{x}) = \frac{p(\boldsymbol{x},\boldsymbol{y})}{p(\boldsymbol{x})} \tag{C.13}$$

Using the sum and product rules of probability [Bis06], $p(y|x)$ becomes:

$$\frac{p(x,y)}{p(x)} = \frac{p(x,y)}{\sum_{y'} p(y',x)}. \tag{C.14}$$

By applying the Hammersley-Clifford's theorem [Cli90] to Conditional Random Fields, the equations are factorized as:

$$\frac{p(y,x)}{\sum_{y'} p(y',x)} = \frac{\prod_{c \in clique(G)} \psi_c(x_c, y_c)}{\sum_{y'} \prod_{c \in clique(G)} \psi_c(x_c, y'_c)} \tag{C.15}$$

The final general model formulation for Conditional Random Fields can be then derived as:

$$p(y|x) = \frac{1}{Z(x)} \prod_{c \in clique(G)} \psi_c(x_c, y_c),\ Z(x) = \sum_{y'} \prod_{c \in clique(G)} \psi_c(x_c, y') \tag{C.16}$$

A CRF is expressed as a log-linear model, which is easier to solve and estimate in statistics. The log-linearity in a CRF occurs in its potential functions. The positive condition of the Hammersley-Clifford theorem [Cli90]: $p(x_1, \ldots, x_n) > 0$ is always true if the factors of equation C.16 are always greater than zero, which means that $\psi(x,y) > 0$. The potential function, also called the factor function, can be therefore written as:

$$\psi_C(x_C, y_C) = \exp\left\{\sum_k (\lambda_{Ck} f_{Ck}(x_C, y_C))\right\} \tag{C.17}$$

for some parameter vector λ_C and some set of feature functions f_{Ck}.

This book makes use of linear-chain Conditional Random Fields only (see Figure C.2, instead of more general CRFs models [SM06]. Equation C.16 can therefore be modified to describe the output variables as a sequence:

$$p(y|x) = \frac{1}{Z(x)} \prod_{j=1}^{n} \psi_j(x,y),\ Z(x) = \sum_{y'} \prod_{j=1}^{n} \psi_j(x,y') \tag{C.18}$$

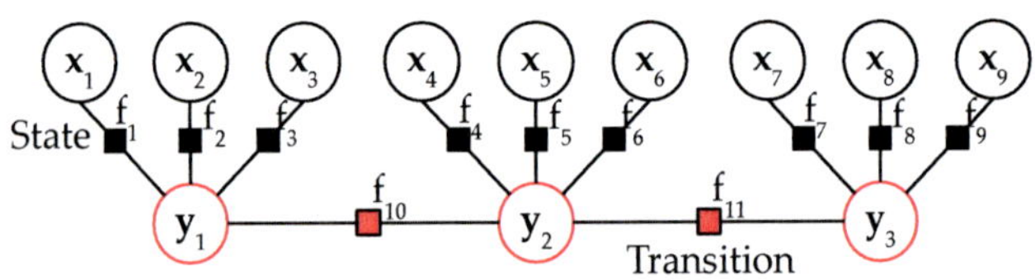

Fig. C.2 A linear-chain Conditional Random Field example. The input nodes x_i are connected via undirected edges (links) to the prediction nodes y_j. The two types of squares represent the nodes for each factor function f_s for state (y_j, x_i) and transition (y_j, y_{j-1}).

The factor functions ψ become:

$$\psi_j(\boldsymbol{x},\boldsymbol{y}) = \exp\left\{\sum_{i=1}^{m} \lambda_i f_i(\boldsymbol{y}_{j-1},\boldsymbol{y}_j,\boldsymbol{x},j)\right\} \tag{C.19}$$

The model of a linear-chain CRF can be written as:

$$\begin{aligned} p_\lambda(\boldsymbol{x}|\boldsymbol{y}) &= \frac{1}{Z_\lambda(\boldsymbol{x})}\prod_{j=1}^{n}\exp\left\{\sum_{i=1}^{m}\lambda_i f_i(\boldsymbol{y}_{i-1},\boldsymbol{y}_i,\boldsymbol{x},j)\right\} \\ &= \frac{1}{Z_\lambda(\boldsymbol{x})}\exp\left\{\sum_{j=1}^{n}\sum_{i=1}^{m}\lambda_i f_i(\boldsymbol{y}_{i-1},\boldsymbol{y}_i,\boldsymbol{x},j)\right\} \end{aligned} \tag{C.20}$$

where the normalization is given by:

$$Z_\lambda(\boldsymbol{x}) = \sum_{\boldsymbol{y}'\in\mathcal{Y}}\exp\left\{\sum_{j=1}^{n}\sum_{i=1}^{m}\lambda_i f_i(\boldsymbol{y}'_{i-1},\boldsymbol{y}'_i,\boldsymbol{x},j)\right\} \tag{C.21}$$

Figure C.2 shows a graphical representation for a linear-chain CRF example. The input nodes $\boldsymbol{x}_i$ are shown with black circles and are connected via undirected links to the prediction nodes $\boldsymbol{y}_j$, shown with red circles. The nodes corresponding to the factor functions f_s are shown with black squares, corresponding to the state $\boldsymbol{y}_j,\boldsymbol{x}_i$ cliques, and with red squares corresponding to the transition cliques $\boldsymbol{y}_j,\boldsymbol{y}_{j-1}$. Each function has its own weight λ. Equation C.20 simply takes the sum of the weights associated with all the squares in the graph, and converts the sum into a probability by normalizing the score. Using the dual state-transition formulation, the potential functions ψ_c can be split into:

- state feature functions:

$$\psi_j(\boldsymbol{x},\boldsymbol{y}_j) = \exp\left\{\sum_{i=1}^{m}\lambda_i s_i(\boldsymbol{y}_j,\boldsymbol{x},j)\right\} \tag{C.22}$$

- transition feature functions:

$$\psi_j(\boldsymbol{y}_j,\boldsymbol{y}_{j-1}) = \exp\left\{\sum_{i=1}^{m}\lambda_i t_i(\boldsymbol{y}_j,\boldsymbol{y}_{j-1})\right\} \tag{C.23}$$

Their combination in the linear-chain CRF equation results in:

$$p_\lambda(\boldsymbol{x}|\boldsymbol{y}) = \frac{1}{Z_\lambda(\boldsymbol{x})}\prod_{j=1}^{n}\exp\left\{\sum_{i=1}^{k}\lambda_i s_i(\boldsymbol{y}_j,\boldsymbol{x},j) + \sum_{i=1}^{l}\lambda_i t_i(\boldsymbol{y}_j,\boldsymbol{y}_{j-1})\right\} \tag{C.24}$$

Learning in a Conditional Random Field is performed by estimating the weights $\lambda_i = \{\lambda_i^1,\ldots,\lambda_i^L\}$. The input consists of a set of training data $\mathcal{T} = \boldsymbol{x}_i$ from a sequence of observation variables, and the output variables sequence $\boldsymbol{y}_j$ of the desired predictions. The parameter estimation is typically performed in a maximum-likelihood fashion. To find good values for the weights λ, the log-likelihood $\mathcal{L}$ on the training data $\mathcal{T}$ needs to be maximized:

$$\mathcal{L}(\mathcal{T}) = \sum_{(\mathbf{x},\mathbf{y})\in\mathcal{T}} \log p(\mathbf{x},\mathbf{y}) \tag{C.25}$$

The log-likelihood and its derivation represent concave functions, and therefore there exists only one global maximum, which guarantees that an optimal solution can be found. A fast way of solving this nonlinear optimization problem is to apply an optimized variant of the Broyden-Fletcher-Goldfarb-Shannon (BFGS) algorithm, namely the Limited-Memory-BFGS method. The L-BFGS method is a derivation of Newton's method and belongs to the second-order class of optimization techniques. Instead of using the Hessian matrix, which has a quadratic size in the number of parameters, at each optimization iteration, the L-BFGS method iteratively finds a minimizer by approximating the inverse of the Hessian by making use of the information from the last m iterations. This optimization saves memory storage and computation time, which helps greatly in dealing with large-scaled problems. The partition function $Z(\boldsymbol{x})$ in the likelihood and the marginal distributions in the gradient can both be computed by using the Forward-Backward Algorithm, which has a computational complexity of $O(|S|^2 n)$, where $\mathcal{S}$ is the number of states and n is the length of the sequence. Each training instance however will have a different partition function and marginals, so the Forward-Backward method needs to be run for each training instance $(\boldsymbol{x},\boldsymbol{y}) \in \mathcal{T}$ at each gradient computation, for a total training cost of $O(|S|^2 n|\mathcal{T}||\mathcal{G}|)$, where $|\mathcal{T}|$ represents the number of training instances and $|\mathcal{G}|$ the number of gradient computations required by the L-BFGS method [SM06].

The inference problem in a CRF is to label an unseen instance $\boldsymbol{y}$ for a certain given observation $\boldsymbol{x}$. A solution is to compute the most-likely sequence using the Viterbi Algorithm [Vit67]. The main difference between the Viterbi method and the Forward-Backward algorithm is that instead of summing, a maximization is applied. The Viterbi algorithm searches this space efficiently with a computational cost that grows only linear in the length of the path.

References

[AA04] Alexa, M., Adamson, A.: On Normals and Projection Operators for Surfaces defined by Point Sets. In: Proceedings of Symposium on Point-Based Graphics 2004, pp. 149–155 (2004)

[AA07] Tsotsos, J.K., Andreopoulos, A.: A Framework for Door Localization and Door Opening Using a Robotic Wheelchair for People Living with Mobility Impairments. In: Robotics: Science and Systems, Workshop: Robot Manipulation: Sensing and Adapting to the Real World, Atlanta, USA (June 30, 2007)

[AAvd07] Alex, F., Anthony, D., van den Anton, H.: Thrift: Local 3D Structure Recognition. In: DICTA 2007: Proceedings of the 9th Biennial Conference of the Australian Pattern Recognition Society on Digital Image Computing Techniques and Applications (2007)

[ABCO+03] Alexa, M., Behr, J., Cohen-Or, D., Fleishman, S., Levin, D., Silva, C.T.: Computing and Rendering Point Set Surfaces. IEEE Transactions on Visualization and Computer Graphics 9(1), 3–15 (2003)

[AK06] Agrawal, M., Konolige, K.: Real-time localization in outdoor environments using stereo vision and inexpensive GPS. In: International Conference on Pattern Recognition (ICPR), Hong Kong, August 20-24 (2006)

[AKB08] Agrawal, M., Konolige, K., Blas, M.R.: CenSurE: Center Surround Extremas for Realtime Feature Detection and Matching. In: Forsyth, D., Torr, P., Zisserman, A. (eds.) ECCV 2008, Part IV. LNCS, vol. 5305, pp. 102–115. Springer, Heidelberg (2008)

[ALAM06] Aude, E.P.L., Lopes, E.P., Aguiar, C.S., Martins, M.F.: Door Crossing and State Identification Using Robotic Vision. In: 8th International IFAC Symposium on Robot Control (Syroco), Bologna, Italy, September 6-8 (2006)

[AM93] Arya, S., Mount, D.M.: Algorithms for Fast Vector Quantization. In: Proceedings of Data Compression Conference, pp. 381–390 (1993)

[AMN+98] Arya, S., Mount, D.M., Netanyahu, N.S., Silverman, R., Wu, A.Y.: An optimal algorithm for approximate nearest neighbor searching fixed dimensions. Journal of the ACM 45(6), 891–923 (1998)

[And07] Nüchter, A., Lingemann, K., Hertzberg, J.: Cached k-d tree search for ICP algorithms. In: Proceedings of the 6th IEEE International Conference on Recent Advances in 3D Digital Imaging and Modeling (3DIM), Montreal, Canada, August 21-23, pp. 419–426. IEEE Computer Society Press (2007)

[ARA+06] Asfour, T., Regenstein, K., Azad, P., Schroeder, J., Bierbaum, A., Vahrenkamp, N., Dillmann, R.: ARMAR-III: An Integrated Humanoid Platform for Sensory-Motor Control. In: IEEE-RAS International Conference on Humanoid Robots (Humanoids), Genoa, Italy, December 4-6 (2006)

[AT08] Andreopoulos, A., Tsotsos, J.K.: Active Vision for Door Localization and Door Opening using Playbot: A Computer Controlled Wheelchair for People with Mobility Impairments. In: Canadian Conference on Computer and Robot Vision, CRV 2008, pp. 3–10 (May 2008)

[AU07] Akagunduz, E., Ulusoy, I.: 3D Object Representation Using Transform and Scale Invariant 3D Features. In: IEEE 11th International Conference on Computer Vision, ICCV 2007, October 14-21 (2007)

[BBK+07] Beetz, M., Bandouch, J., Kirsch, A., Maldonado, A., Müller, A., Rusu, R.B.: The assistive kitchen — a demonstration scenario for cognitive technical systems. In: Proceedings of the 4th COE Workshop on Human Adaptive Mechatronics, HAM (2007)

[BBL09] Bae, K.-H., Belton, D., Lichti, D.D.: A Closed-Form Expression of the Positional Uncertainty for 3D Point Clouds. IEEE Transactions on Pattern Analysis and Machine Intelligence 31(4), 577–590 (2009)

[BC94] Berkmann, J., Caelli, T.: Computation of Surface Geometry and Segmentation Using Covariance Techniques. IEEE Transactions on Pattern Analysis and Machine Intelligence (PAMI) 16(11), 1114–1116 (1994)

[BETG08] Bay, H., Ess, A., Tuytelaars, T., Van Gool, L.: SURF: Speeded Up Robust Features. Computer Vision and Image Understanding (CVIU) 110(3), 346–359 (2008)

[BGV92] Bernhard, E., Boser, I.M.: Guyon, and Vladimir N. Vapnik. A Training Algorithm for Optimal Margin Classifiers. In: Proceedings of the 5th Annual ACM Workshop on Computational Learning Theory, pp. 144–152. ACM Press (1992)

[BH95] Burel, G., Hénocq, H.: Three-dimensional invariants and their application to object recognition. Signal Process 45(1), 1–22 (1995)

[Bis06] Bishop, C.M.: Pattern Recognition and Machine Learning (Information Science and Statistics). Springer (August 2006) ISBN 0387310738

[BL04] Bae, K.-H., Lichti, D.D.: Automated registration of unorganized point clouds from terrestrial laser scanners. In: International Archives of Photogrammetry and Remote Sensing (IAPRS), pp. 222–227 (2004)

[BM92] Besl, P.J., McKay, N.D.: A Method for Registration of 3-D Shapes. IEEE Trans. Pattern Anal. Mach. Intell. 14 (1992)

[BOV03] Barla, A., Odone, F., Verri, A.: Histogram intersection kernel for image classification. In: Proceedings of International Conference on Image Processing (ICIP), Barcelona, Catalonia, Spain, September 14-18 (2003)

[BOW+07] Borst, C., Ott, C., Wimbock, T., Brunner, B., Zacharias, F., Baeum, B., Hillenbrand, U., Haddadin, S., Albu-Schaeffer, A., Hirzinger, G.: A humanoid upper body system for two-handed manipulation. In: IEEE International Conference on Robotics and Automation (ICRA), Rome, Italy, April 10-14 (2007)

[BP06] Del Bimbo, A., Pala, P.: Content-based retrieval of 3D models. ACM Trans. Multimedia Comput. Commun. Appl. 2(1), 20–43 (2006)

[BRBB09] Blas, M.R., Rusu, R.B., Blanke, M., Beetz, M.: Fault-tolerant 3D Mapping with Application to an Orchard Robot. In: Proceedings of the 7th IFAC International Symposium on Fault Detection, Supervision and Safety of Technical Processes (SAFEPROCESS), Barcelona, Spain, June 30 - July 3 (2009)

[Bur98] Christopher, J.C.: Burges. A tutorial on support vector machines for pattern recognition. Data Mining and Knowledge Discovery 2, 121–167 (1998)

[CF] Campbell, R.J., Flynn, P.J.: Eigenshapes for 3D object recognition in range data. In: IEEE Computer Society Conference on Computer Vision and Pattern Recognition, pp. 505–510 (1999)

[CHV99] Chapelle, O., Haffner, P., Vapnik, V.N.: Support vector machines for histogram-based image classification. IEEE Transactions on Neural Networks 10(5), 1055–1064 (1999)

[CLH$^+$05] Choset, H., Lynch, K.M., Hutchinson, S., Kantor, G.A., Burgard, W., Kavraki, L.E., Thrun, S.: Principles of Robot Motion: Theory, Algorithms, and Implementations. MIT Press (June 2005)

[Cli90] Clifford, P.: Markov Random Fields in statistics. In: Disorder in Physical Systems, pp. 19–32. Oxford University Press (1990)

[CM02] Chum, O., Matas, J.: Randomized RANSAC with Td,d test. In: Image and Vision Computing (IVCNZ), Auckland, New Zealand, November 26-28, pp. 448–457 (2002)

[Con09] Computing Community Consortium. A Roadmap for US Robotics: From Internet to Robotics. Technical report (May 21, 2009)

[CV95] Cortes, C., Vapnik, V.: Support-Vector Networks. Machine Learning, 273–297 (1995)

[DHKL01] Dyn, N., Hormann, K., Kim, S.-J., Levin, D.: Optimizing 3D triangulations using discrete curvature analysis. In: Mathematical Methods for Curves and Surfaces (Oslo 2000), pp. 135–146. Vanderbilt University, Nashville (2001)

[DHS00] Duda, R.O., Hart, P.E., Stork, D.G.: Pattern Classification, 2nd edn. Wiley-Interscience (November 2000) ISBN 0471056693

[DK08] Diankov, R., Kuffner, J.: OpenRAVE: A Planning Architecture for Autonomous Robotics. Technical Report CMU-RI-TR-08-34, Robotics Institute, Carnegie Mellon University (July 2008)

[EJL07] Sierra, B., Jauregi, E., Martinez-Otzeta, J.M., Lazkano, E.: Door Handle Identification: a Three-Stage approach. In: 6th IFAC Symposium on Intelligent Autonomous Vehicles (IAV), Toulouse, France, September 3-5 (2007)

[EK08] Ng Ellen Klingbeil, A.Y., Saxena, A.: Learning to Open New Doors. In: Robotics Science and Systems (RSS) workshop on Robot Manipulation, Zurich, Switzerland (June 28, 2008)

[Elf89] Elfes, A.: Using Occupancy Grids for Mobile Robot Perception and Navigation. IEEE Computer 22(6), 46–57 (1989)

[ESN06] Engels, C., Stewenius, H., Nister, D.: Bundle Adjustment Rules. Photogrammetric Computer Vision (September 2006)

[FB81] Martin Fischler, A., Robert Bolles, C.: Random sample consensus: a paradigm for model fitting with applications to image analysis and automated cartography. Communications of the ACM 24(6), 381–395 (1981)

[FCOS05] Fleishman, S., Cohen-Or, D., Silva, C.T.: Robust moving least-squares fitting with sharp features. In: SIGGRAPH 2005: ACM SIGGRAPH 2005 Papers, pp. 544–552 (2005)

[Fit01] Fitzgibbon, A.W.: Robust Registration of 2D and 3D Point Sets. In: Proceedings of the British Machine Vision Conference, pp. 662–670 (2001)

[FKaLW98] Frey, B.J., Kschischang, F.R., Loeliger, H.A., Wiberg, N.: Factor Graphs and Algorithms. In: Proc. 35th Allerton Conf. on Communications, Control, and Computing, Allerton, pp. 666–680 (1998)

[GA05] Gruen, A., Akca, D.: Least squares 3D surface and curve matching. International Journal of Photogrammetry and Remote Sensing 59, 151–174 (2005)

[GGGZ05] Gatzke, T., Grimm, C., Garland, M., Zelinka, S.: Curvature Maps for Local Shape Comparison. In: SMI 2005: Proceedings of the International Conference on Shape Modeling and Applications 2005 (SMI 2005), pp. 246–255 (2005)

[GIRL03] Gelfand, N., Ikemoto, L., Rusinkiewicz, S., Levoy, M.: Geometrically Stable Sampling for the ICP Algorithm. In: Proceedings of the 4th IEEE International Conference on Recent Advances in 3D Digital Imaging and Modeling (3DIM), Banff, Canada, October 6-10 (2003)

[GKS00] Gopi, M., Krishnan, S., Silva, C.T.: Surface reconstruction based on lower dimensional localized delaunay triangulation. In: Gross, M., Hopgood, F.R.A. (eds.) Computer Graphics Forum (Eurographics), vol. 19(3) (2000)

[GMGP05] Gelfand, N., Mitra, N.J., Guibas, L.J., Pottmann, H.: Robust Global Registration. In: Proceedings of the Symposium of Geometric Processing (2005)

[GvL96] Golub, G.H., van Loan, C.F.: Matrix Computations, 3rd edn. Johns Hopkins University Press (1996)

[GWM01] Gumhold, S., Wang, X., MacLeod, R.: Features Extraction from Point Clouds. In: Proceedings of the 10th International Meshing Roundtable, pp. 293–305 (2001)

[HBG06] Bay, H., Tuytelaars, T., Van Gool, L.: SURF: Speeded Up Robust Features. In: Leonardis, A., Bischof, H., Pinz, A. (eds.) ECCV 2006. LNCS, vol. 3951, pp. 404–417. Springer, Heidelberg (2006)

[HDD+92] Hoppe, H., DeRose, T., Duchamp, T., McDonald, J., Stuetzle, W.: Surface reconstruction from unorganized points. In: Proceedings of the 19th Annual Conference on Computer Graphics and Interactive Techniques (SIGGRAPH), Chicago, Illinois, USA, July 27-31, pp. 71–78 (1992)

[HEH05] Hoiem, D., Efros, A.A., Hebert, M.: Geometric context from a single image. In: International Conference on Computer Vision (ICCV), Beijing, China, October 17-20, pp. 654–661 (2005)

[HKDH04] Huber, D., Kapuria, A., Donamukkala, R.R., Hebert, M.: Parts-based 3D object classification. In: Proceedings of the IEEE Conference on Computer Vision and Pattern Recognition, CVPR 2004 (June 2004)

[HLLS01] Hetzel, G., Leibe, B., Levi, P., Schiele, B.: 3D Object Recognition from Range Images using Local Feature Histograms. In: IEEE Computer Society Conference on Computer Vision and Pattern Recognition, vol. 2, p. 394 (2001)

[Hor84] Horn, B.K.P.: Extended Gaussian Images. Proceedings of the IEEE 72, 1671–1686 (1984)

[HZ04] Hartley, R.I., Zisserman, A.: Multiple View Geometry in Computer Vision, 2nd edn. Cambridge University Press (2004) ISBN 0521540518

[ItSMaUoT07] IRT (Information, Robot Technology) Foundation to Support Man, and Aging Society at University of Tokyo. Mission Statement. Technical report (2007)

[JD00] Jain, A.K., Dorai, C.: 3D object recognition: Representation and matching. Statistics and Computing 10(2), 167–182 (2000)

[JH99] Johnson, A., Hebert, M.: Using spin images for efficient object recognition in cluttered 3D scenes. IEEE Transactions on Pattern Analysis and Machine Intelligence 21(5), 433–449 (1999)

[JH03] Jost, J., Hugli, H.: A Multi-Resolution ICP with Heuristic Closest Point Search for Fast and Robust 3D Registration of Range Images. 3dim 00, 427 (2003)

[JK97] Johnson, A.E., Kang, S.B.: Registration and integration of textured 3-D data. In: NRC 1997: Proceedings of the International Conference on Recent Advances in 3-D Digital Imaging and Modeling, p. 234 (1997)

[JNY07] Jiang, Y.-G., Ngo, C.-W., Yang, J.: Towards optimal bag-of-features for object categorization and semantic video retrieval. In: CIVR 2007: Proceedings of the 6th ACM International Conference on Image and Video Retrieval, pp. 494–501 (2007)

[KAI05] Konolige, K., Agrawal, M., Iocchi, L.: Real-time detection of independent motion using stereo. In: IEEE Workshop on Motion (WACV/MOTION), Breckenridge, CO, USA, January 5-7 (2005)

[KAWB09] Klasing, K., Althoff, D., Wollherr, D., Buss, M.: Comparison of Surface Normal Estimation Methods for Range Sensing Applications. In: Proceedings of the IEEE International Conference on Robotics and Automation (ICRA), Kobe, Japan, May 12-17 (2009)

[KHY+06] Katz, D., Horrell, E., Yang, Y., Burns, B., Buckley, T., Grishkan, A., Zhylkovskyy, V., Brock, O., Learned-Miller, E.: The UMass Mobile Manipulator UMan: An Experimental Platform for Autonomous Mobile Manipulation. In: Robotics Science and Systems (RSS) Workshop Manipulation for Human Environments, Philadelphia, USA, August 19 (2006)

[Kon97] Konolige, K.: Small Vision Systems: Hardware and Implementation. In: Eighth International Symposium on Robotics Research, Hayama, Japan (October 1997)

[KSNS07] Kalogerakis, E., Simari, P., Nowrouzezahrai, D., Singh, K.: Robust statistical estimation of curvature on discretized surfaces. In: SGP 2007: Proceedings of the Fifth Eurographics Symposium on Geometry Processing, pp. 13–22 (2007)

[KvD92] Koenderink, J.J., van Doorn, A.J.: Surface shape and curvature scales. Image and Vision Computing 10(8), 557–565 (1992)

[LaV06] LaValle, S.M.: Planning Algorithms. Cambridge University Press, Cambridge (2006), `http://planning.cs.uiuc.edu/`

[LEF95] Lorusso, A., Eggert, D., Fisher, R.: A comparison of four algorithms for estimating 3-d rigid transformations (1995)

[Lev03] Levin, D.: Mesh-independent surface interpolation. In: Brunnett, H., Mueller (eds.) Geometric Modeling for Scientific Visualization, pp. 37–49. Springer, Heidelberg (2003)

[Lin00] Lindstrom, P.: Out-of-core simplification of large polygonal models. In: Proceedings of the 27th Annual Conference on Computer Graphics and Interactive Techniques (SIGGRAPH), pp. 259–262. ACM Press/Addison-Wesley Publishing Co., New York (2000)

[LMP01] Lafferty, J., Mccallum, A., Pereira, F.: Conditional Random Fields: Probabilistic Models for Segmenting and Labeling Sequence Data. In: Proceedings of the 18th International Conference on Machine Learning, Williamstown, MA, USA, June 28 - July 1 (2001)

[Low04a] Lowe, D.G.: Distinctive image features from scale-invariant keypoints. International Journal of Computer Vision 60(2), 91–110 (2004)

[Low04b] Lowe, D.G.: Distinctive Image Features from Scale-Invariant Keypoints. Int. J. Comput. Vision 60(2), 91–110 (2004)

[LUVH05] Lalonde, J.-F., Unnikrishnan, R., Vandapel, N., Hebert, M.: Scale Selection for Classification of Point-sampled 3-D Surfaces. In: Proceedings of the 5th International Conference on 3-D Digital Imaging and Modeling (3DIM), Ottawa, Ontario, Canada, June 13-16, pp. 285–292 (2005)

[MA04] Miller, A., Allen, P.K.: Graspit!: A Versatile Simulator for Robotic Grasping. IEEE Robotics and Automation Magazine 11(4), 110–122 (2004)

[Mar63] Marquardt, D.W.: An algorithm for least-squares estimation of nonlinear parameters. SIAM Journal on Applied Mathematics 11(2), 431–441 (1963)

[MC08] Ben-Chen, M., Gotsman, C.: Characterizing shape using conformal factors. In: Eurographics Workshop on 3D Object Retrieval (2008)

[MDH+08] May, S., Droeschel, D., Holz, D., Wiesen, C., Fuchs, S.: 3d pose estimation and mapping with time-of-flight cameras (September 26, 2008)

[MGP+04] Mitra, N.J., Gelfand, N., Pottmann, H., Guibas, L.: Registration of point cloud data from a geometric optimization perspective. In: Scopigno, R., Zorin, D. (eds.) Eurographics Symposium on Geometry Processing, pp. 23–32 (2004)

[Mit97] Mitchell, T.M.: Machine Learning. McGraw-Hill, New York (1997)

[MK08] Inaba, M., Kojima, M., Okada, K.: Manipulation and Recognition of Objects Incorporating Joints by a Humanoid Robot for Daily Assistive Tasks. In: Proceedings of the IEEE/RSJ International Conference on Intelligent Robots and Systems (IROS), Nice, France, September 22-26 (2008)

[ML09] Muja, M., Lowe, D.G.: Fast Approximate Nearest Neighbors with Automatic Algorithm Configuration. In: Proceedings of the International Conference on Computer Vision Theory and Applications (VISAPP), Lisbon, Portugal, February 5-8 (2009)

[MN03] Mitra, N.J., Nguyen, A.: Estimating surface normals in noisy point cloud data. In: Proceedings of The 19th ACM Symposium on Computational Geometry (SCG), San Diego, California, USA, June 8-10, pp. 322–328 (2003)

[MPD06] Makadia, A., Patterson IV, A., Daniilidis, K.: Fully Automatic Registration of 3D Point Clouds. In: CVPR 2006: Proceedings of the 2006 IEEE Computer Society Conference on Computer Vision and Pattern Recognition, pp. 1297–1304 (2006)

[MRB09] Marton, Z.C., Rusu, R.B., Beetz, M.: On Fast Surface Reconstruction Methods for Large and Noisy Datasets. In: Proceedings of the IEEE International Conference on Robotics and Automation (ICRA), Kobe, Japan, May 12-17 (2009)

[MRJ+09] Marton, Z.C., Rusu, R.B., Jain, D., Klank, U., Beetz, M.: Probabilistic Categorization of Kitchen Objects in Table Settings with a Composite Sensor. In: Proceedings of the IEEE/RSJ International Conference on Intelligent Robots and Systems (IROS), St. Louis, MO, USA, October 11-15 (2009)

[MRS+09] Morisset, B., Rusu, R.B., Sundaresan, A., Hauser, K., Agrawal, M., Latombe, J.-C., Beetz, M.: Leaving Flatland: Toward Real-Time 3D Navigation. In: Proceedings of the IEEE International Conference on Robotics and Automation (ICRA), Kobe, Japan, May 12-17 (2009)

[MTJ+07] Mozos, O.M., Triebel, R., Jensfelt, P., Rottmann, A., Burgard, W.: Supervised Semantic Labeling of Places using Information Extracted from Laser and Vision Sensor Data. Robotics and Autonomous Systems 55(5), 391–402 (2007)

[New56] Newman, J.R.: The World of Mathematics. Dover, New York (1956) (2003) ISBN 0486432688

[NH08] Nuechter, A., Hertzberg, J.: Towards Semantic Maps for Mobile Robots. Journal of Robotics and Autonomous Systems (JRAS), Special Issue on Semantic Knowledge in Robotics 56, 915–926 (2008)

[NJAK08] Nguyen, H., Jain, A., Anderson, C., Kemp, C.C.: A Clickable World: Behavior Selection Through Pointing and Context for Mobile Manipulation. In: IEEE/RJS International Conference on Intelligent Robots and Systems (IROS), Nice, France, September 22-26 (2008)

[NSG08] Nalpantidis, L., Sirakoulis, G.C., Gasteratos, A.: Review of Stereo Vision Algorithms: from Software to Hardware. International Journal of Optomechatronics 2(4), 435–462 (2008)

[NSH03] Nuechter, A., Surmann, H., Hertzberg, J.: Automatic Model Refinement for 3D Reconstruction with Mobile Robots. In: Proceedings of the 4th IEEE International Conference on Recent Advances in 3D Digital Imaging and Modeling (3DIM), Banff, Canada, October 6-10 (2003)

[Nue09] Nuechter, A.: 3D Robotic Mapping. In: Springer Tracts in Advanced Robotics, vol. 52, Springer, Heidelberg (2009)

[NWL+05] Nüchter, A., Wulf, O., Lingemann, K., Hertzberg, J., Wagner, B., Surmann, H.: 3D Mapping with Semantic Knowledge. In: Bredenfeld, A., Jacoff, A., Noda, I., Takahashi, Y. (eds.) RoboCup 2005. LNCS (LNAI), vol. 4020, pp. 335–346. Springer, Heidelberg (2006)

[OBBH07] Ott, C., Baeuml, B., Borst, C., Hirzinger, G.: Autonomous Opening of a Door with a Mobile Manipulator: A Case Study. In: 6th IFAC Symposium on Intelligent Autonomous Vehicles (IAV), Toulouse, France, September 3-5 (2007)

[OFCD02] Osada, R., Funkhouser, T., Chazelle, B., Dobkin, D.: Shape distributions. ACM Transactions on Graphics 21, 807–832 (2002)

[OLK+04] Oggier, T., Lehmann, M., Kaufmann, R., Schweizer, M., Richter, M., Metzler, P., Lang, G., Lustenberger, F., Blanc, N.: An all-solid-state optical range camera for 3D real-time imaging with sub-centimeter depth resolution (SwissRanger). Optical Design and Engineering 5249(1), 534–545 (2004)

[PCF05] Paragios, N., Chen, Y., Faugeras, O.: Handbook of Mathematical Models in Computer Vision. Springer-Verlag New York, Inc. (2005) ISBN 0387263713

[Pet02] Petitjean, S.: A survey of methods for recovering quadrics in triangle meshes. ACM Computing Surveys 34(2), 211–262 (2002)

[PGK02] Pauly, M., Gross, M., Kobbelt, L.: Efficient Simplification of Point-Sampled Surfaces. In: Proceedings of IEEE Visualization, Boston, MA, USA, October 27-November 01 (2002)

[PKG03] Pauly, M., Keiser, R., Gross, M.: Multi-scale Feature Extraction on Point-Sampled Surfaces. Computer Graphics Forum 22(3), 281–289 (2003)

[Pla09] European Robotics Technology Platform. Robotic Visions To 2020 And Beyond – The Strategic Research Agenda For Robotics in Europe. Technical report (June 2009)

[PLH04] Pottmann, H., Leopoldseder, S., Hofer, M.: Registration without ICP. Computer Vision and Image Understanding 95(1), 54–71 (2004)

[PLT07] Phillips, J.M., Liu, R., Tomasi, C.: Outlier Robust ICP for Minimizing Fractional RMSD. 3dim, 427–434 (2007)

[PRLLM08] de la Puente, P., Rodríguez-Losada, D., López, R., Matía, F.: Extraction of Geometrical Features in 3D Environments for Service Robotic Applications. In: Corchado, E., Abraham, A., Pedrycz, W. (eds.) HAIS 2008. LNCS (LNAI), vol. 5271, pp. 441–450. Springer, Heidelberg (2008)

[PSK+02] Page, D.L., Sun, Y., Koschan, A.F., Paik, J., Abidi, M.A.: Normal vector voting: crease detection and curvature estimation on large, noisy meshes. Graph. Models 64(3/4), 199–229 (2002)

[PSM07] Paul, S., Saad, A., Mubarak, S.: A 3-dimensional sift descriptor and its application to action recognition. In: MULTIMEDIA 2007: Proceedings of the 15th International Conference on Multimedia (2007)

[QBN07] Quigley, M., Berger, E., Ng, A.Y.: STAIR: Hardware and Software Architecture. In: AAAI 2007 Mobile Robot Workshop, Vancouver, BC, Canada (July 23, 2007)

[RB05] Ruto, A., Buxton, B.: Application of a Robust and Efficient ICP Algorithm for Fitting a Deformable 3D Human Torso Model to Noisy Data. In: DICTA 2005: Proceedings of the Digital Image Computing on Techniques and Applications, vol. 62 (2005)

[RBB09] Rusu, R.B., Blodow, N., Beetz, M.: Fast Point Feature Histograms (FPFH) for 3D Registration. In: Proceedings of the IEEE International Conference on Robotics and Automation (ICRA), Kobe, Japan, May 12-17 (2009)

[RBM+07] Rusu, R.B., Blodow, N., Marton, Z.-C., Soos, A., Beetz, M.: Towards 3d object maps for autonomous household robots. In: Proceedings of the 20th IEEE International Conference on Intelligent Robots and Systems (IROS), San Diego, CA, USA, October 29-November 2 (2007)

[RBM+09] Rusu, R.B., Bandouch, J., Meier, F., Essa, I., Beetz, M.: Human Action Recognition using Global Point Feature Histograms and Action Shapes. Advanced Robotics Journal, Robotics Society of Japan, RSJ (2009)

[RBMB08] Rusu, R.B., Blodow, N., Marton, Z.C., Beetz, M.: Aligning Point Cloud Views using Persistent Feature Histograms. In: Proceedings of the 21st IEEE/RSJ International Conference on Intelligent Robots and Systems (IROS), Nice, France, September 22-26 (2008)

[RBMB09] Rusu, R.B., Blodow, N., Marton, Z.C., Beetz, M.: Close-range Scene Segmentation and Reconstruction of 3D Point Cloud Maps for Mobile Manipulation in Human Environments. In: Proceedings of the IEEE/RSJ International Conference on Intelligent Robots and Systems (IROS), St. Louis, MO, USA, October 11-15 (2009)

[RHBB09] Rusu, R.B., Holzbach, A., Blodow, N., Beetz, M.: Fast Geometric Point Labeling using Conditional Random Fields. In: Proceedings of the IEEE/RSJ International Conference on Intelligent Robots and Systems (IROS), St. Louis, MO, USA, October 11-15 (2009)

[RL01] Rusinkiewicz, S., Levoy, M.: Efficient variants of the ICP algorithm. In: Proceedings of the Third International Conference on 3-D Digital Imaging and Modeling, pp. 145–152 (2001)

[RMB+08a] Rusu, R.B., Marton, Z.C., Blodow, N., Dolha, M., Beetz, M.: Towards 3D Point Cloud Based Object Maps for Household Environments. Robotics and Autonomous Systems Journal (Special Issue on Semantic Knowledge) (2008)

[RMB+08b] Rusu, R.B., Marton, Z.C., Blodow, N., Dolha, M.E., Beetz, M.: Functional Object Mapping of Kitchen Environments. In: Proceedings of the 21st IEEE/RSJ International Conference on Intelligent Robots and Systems (IROS), Nice, France, September 22-26 (2008)

[RMBB08a] Rusu, R.B., Marton, Z.C., Blodow, N., Beetz, M.: Interpretation of Urban Scenes based on Geometric Features. In: Proceedings of the 21st IEEE/RSJ International Conference on Intelligent Robots and Systems (IROS) Workshop on 3D Mapping, Nice, France (September 26, 2008) (invited paper)

[RMBB08b] Rusu, R.B., Marton, Z.C., Blodow, N., Beetz, M.: Learning Informative Point Classes for the Acquisition of Object Model Maps. In: Proceedings of the 10th International Conference on Control, Automation, Robotics and Vision (ICARCV), Hanoi, Vietnam, December 17-20 (2008)

[RMBB08c] Rusu, R.B., Marton, Z.C., Blodow, N., Beetz, M.: Persistent Point Feature Histograms for 3D Point Clouds. In: Proceedings of the 10th International Conference on Intelligent Autonomous Systems (IAS-10), Baden-Baden, Germany, July 23-25 (2008)

[RMCB09] Rusu, R.B., Meeussen, W., Chitta, S., Beetz, M.: Laser-based Perception for Door and Handle Identification. In: Proceedings of the International Conference on Advanced Robotics (ICAR), Munich, Germany, June 22-26 (2009)

[Rou84] Rousseeuw, P.J.: Least Median of Squares Regression. Journal of the American Statistical Association (ASAJ) 79, 871–880 (1984)

[RSG+09] Rusu, R.B., Sucan, I.A., Gerkey, B., Chitta, S., Beetz, M., Kavraki, L.E.: Real-time Perception-Guided Motion Planning for a Personal Robot. In: Proceedings of the IEEE/RSJ International Conference on Intelligent Robots and Systems (IROS), St. Louis, MO, USA, October 11-15 (2009)

[RSM+08] Rusu, R.B., Sundaresan, A., Morisset, B., Agrawal, M., Beetz, M.: Leaving Flatland: Realtime 3D Stereo Semantic Reconstruction. In: Proceedings of the International Conference on Intelligent Robotics and Applications (ICIRA), Wuhan, China, October 15-17 (2008)

[RSM+09] Rusu, R.B., Sundaresan, A., Morisset, B., Hauser, K., Agrawal, M., Latombe, J.-C., Beetz, M.: Leaving Flatland: Efficient Real-Time 3D Navigation. International Journal of Field Robotics (Special Issue on 3D Mapping) (2009)

[SA01] Sun, Y., Abidi, M.A.: Surface matching by 3D point's fingerprint. In: Proc. IEEE Int'l Conf. on Computer Vision, vol. II, pp. 263–269 (2001)

[SAH08] Silpa-Anan, C., Hartley, R.: Optimized KD-trees for fast image descriptor matching. In: Proceedings of IEEE Computer Society Conference on Computer Vision and Pattern Recognition, Anchorage, Alaska, June 23-28 (2008)

[SBK01] Saranli, U., Buehler, M., Koditschek, D.E.: RHex: A Simple and Highly Mobile Hexapod Robot. International Journal of Robotics Research 20(1), 616–631 (2001)

[SBW04] Estépar, R.S.J., Brun, A., Westin, C.-F.: Robust Generalized Total Least Squares Iterative Closest Point Registration. In: Barillot, C., Haynor, D.R., Hellier, P. (eds.) MICCAI 2004. LNCS, vol. 3216, pp. 234–241. Springer, Heidelberg (2004)

[Sch07] Schaal, S.: The new robotics - towards human-centered machines. HFSP Journal Frontiers of Interdisciplinary Research in the Life Sciences (2), 115–126 (2007)

[SDW93] Simèon, T., Dacre-Wright, B.: A Practical Motion Planner for All-terrain Mobile Robots. In: IEEE/RSJ International Conference on Intelligent Robots and Systems (IROS), Yokohama, Japan, July 26-30 (1993)

[SFW+08] Srinivasa, S., Ferguson, D., Weghe, M.V., Diankov, R., Berenson, D., Helfrich, C., Strasdat, H.: The Robotic Busboy: Steps Towards Developing a Mobile Robotic Home Assistant. In: International Conference on Intelligent Autonomous Systems (IAS-10), Baden-Baden, Germany, July 23-25 (2008)

[SH80] Sadjadi, F.A., Hall, E.L.: Three-Dimensional Moment Invariants. PAMI 2(2), 127–136 (1980)

[Sha98] Shakarji, C.: Least-Squares Fitting Algorithms of the NIST Algorithm Testing System. Journal of Research of the National Institute of Standards and Technology 103(6), 633–641 (1998)

[SL02] Sánchez, G., Latombe, J.-C.: On Delaying Collision Checking in PRM Planning: Application to Multi-Robot Coordination. International Journal of Robotics Research 21(1), 5–26 (2002)

[SLW02] Sharp, G., Lee, S., Wehe, D.: ICP registration using invariant features. IEEE Transactions on Pattern Analysis and Machine Intelligence (PAMI) 24(1), 90–102 (2002)

[SM06] Sutton, C., McCallum, A.: An Introduction to Conditional Random Fields for Relational Learning. In: Introduction to Statistical Relational Learning, ch. 4, pp. 93–128. MIT Press (2006)

[SMRR07] Soares, L., Ménier, C., Raffin, B., Roch, J.-L.: Parallel Adaptive Octree Carving for Real-time 3D Modeling. In: IEEE Virtual Reality Conference, Charlotte, USA, March 14-17 (2007)

[SS97] Sun, C., Sherrah, J.: 3D Symmetry Detection Using The Extended Gaussian Image. IEEE Transactions on Pattern Analysis and Machine Intelligence 19(2), 164–168 (1997)

[SS07] Skelly, L.J., Sclaroff, S.: Improved feature descriptors for 3-D surface matching. In: Proc. SPIE Conf. on Two- and Three-Dimensional Methods for Inspection and Metrology V (2007)

[SSN08] Saxena, A., Sun, M., Ng, A.Y.: Make3d: Learning 3-d scene structure from a single still image. IEEE Transactions on Pattern Analysis and Machine Intelligence, PAMI (2008)

[STC04] Shawe-Taylor, J., Cristianini, N.: Kernel Methods for Pattern Analysis. Cambridge University Press, New York (2004) ISBN 0521813972

[SWBR04] Schröter, D., Weber, T., Beetz, M., Radig, B.: Detection and Classification of Gateways for the Acquisition of Structured Robot Maps. In: Rasmussen, C.E., Bülthoff, H.H., Schölkopf, B., Giese, M.A. (eds.) DAGM 2004. LNCS, vol. 3175, pp. 553–561. Springer, Heidelberg (2004)

[SWK07a] Schnabel, R., Wahl, R., Klein, R.: Efficient RANSAC for Point-Cloud Shape Detection. Computer Graphics Forum 26(2), 214–226 (2007)

[SWK07b] Schnabel, R., Wahl, R., Klein, R.: Efficient RANSAC for Point-Cloud Shape Detection. Computer Graphics Forum 26(2), 214–226 (2007)

[THA07] Tang, P., Huber, D., Akinci, B.: A Comparative Analysis of Depth-Discontinuity and Mixed-Pixel Detection Algorithms, August 21-23 (2007)

[TL94] Turk, G., Levoy, M.: Zippered Polygon Meshes from Range Images. In: Proceedings of the 21st Annual Conference on Computer Graphics and Interactive Techniques (SIGGRAPH), Orlando, Florida, July 24-29, pp. 311–318 (1994)

[TV04] Tangelder, J.W.H., Veltkamp, R.C.: A Survey of Content Based 3D Shape Retrieval Methods. In: SMI 2004: Proceedings of the Shape Modeling International, pp. 145–156 (2004)

[TZ00] Torr, P., Zisserman, A.: MLESAC: A new robust estimator with application to estimating image geometry. Computer Vision and Image Understanding 78, 138–156 (2000)

[TZG96] Taubin, G., Zhang, T., Golub, G.H.: Optimal Surface Smoothing as Filter Design. In: Buxton, B.F., Cipolla, R. (eds.) ECCV 1996. LNCS, vol. 1065, pp. 283–292. Springer, Heidelberg (1996)

[Ume91] Umeyama, S.: Least-Squares Estimation of Transformation Parameters Between Two Point Patterns. IEEE Trans. Pattern Anal. Mach. Intell. 13(4), 376–380 (1991)

[Unn08] Unnikrishnan, R.: Statistical Approaches to Multi-Scale Point Cloud Processing. PhD thesis, Robotics Institute, Carnegie Mellon University, Pittsburgh, PA (May 2008)

[Vas03] Vasconcelos, N.: Feature Selection by Maximum Marginal Diversity: optimality and implications for visual recognition. In: IEEE Computer Society Conference on Computer Vision and Pattern Recognition (CVPR), Madison, Wisconsin, USA, June 18-20 (2003)

[Vit67] Viterbi, A.J.: Error bounds for convolutional codes and an asymptotically optimum decoding algorithm. IEEE Transactions on Information Theory 13(2), 260–269 (1967)

[WGS03] Weingarten, J., Gruener, G., Siegwart, R.: A Fast and Robust 3D Feature Extraction Algorithm for Structured Environment Reconstruction. In: International Conference on Advanced Robotics (ICAR), Coimbra, Portugal, June 30-July 3 (2003)

[WHH03] Wahl, E., Hillenbrand, U., Hirzinger, G.: Surflet-Pair-Relation Histograms: A Statistical 3D-Shape Representation for Rapid Classification. In: The 4th International Conference on 3-D Digital Imaging and Modeling, Banff, Alberta, Canada, October 6-10 (2003)

[WO03] Wang, J., Oliveira, M.M.: A Hole-Filling Strategy for Reconstruction of Smooth Surfaces in Range Images. Sibgrapi, 11 (2003)

[YLHP06] Yang, Y.-L., Lai, Y.-K., Hu, S.-M., Pottmann, H.: Robust principal curvatures on multiple scales. In: SGP 2006: Proceedings of the Fourth Eurographics Symposium on Geometry Processing, pp. 223–226 (2006)

[Zha] Zhang, Z.: Iterative Point Matching for Registration of Free-Form Curves. Technical Report RR-1658

[ZTCS99] Zhang, R., Tsai, P.-S., Cryer, J.E., Shah, M.: Shape from Shading: A Survey. IEEE Transactions on Pattern Analysis and Machine Intelligence (PAMI) 21(8), 690–706 (1999)